D1293402

The Safety Management Primer

The Safety Management Primer

Leon Schenkelbach, CSP

1975

Dow Jones-Irwin, Inc. Homewood, Illinois 60430

HD	Schenkelbach, Leon.
7262	
.S325	The safety
	management primer
658.382 S3	

© DOW JONES-IRWIN, INC., 1975

All rights reserved. No part of this publication may be
reproduced, stored in a retrieval system, or transmitted,
in any form or by any means, electronic, mechanical,
photocopying, recording, or otherwise, without the prior
written permission of the publisher.

This publication is designed to provide accurate and
authoritative information in regard to the subject matter
covered. It is sold with the understanding that the
publisher is not engaged in rendering legal, accounting, or
other professional service. If legal advice or other expert
assistance is required, the services of a competent
professional person should be sought.

*From a Declaration of Principles jointly adopted by a Committee
of the American Bar Association and a Committee of Publishers.*

First Printing, May 1975

ISBN 0-87094-092-9
Library of Congress Catalog Card No. 74–25816
Printed in the United States of America

DABNEY LANCASTER LIBRARY
LONGWOOD COLLEGE
FARMVILLE, VIRGINIA 23901

To my wife,
LUCILLE
who patiently recommended,
probed, edited, sympathized, and
agonized with me during
the metamorphosis and
creation of this book.

DABNEY LANCASTER LIBRARY

1000180285

75-06411

Foreword

Mr. Schenkelbach's excellent "primer" on safety practices and theories as they apply to industry should be required reading from the Office of the President to the foreman on the line. It is both a reminder to practice what we already know as well as a provocative discussion of new thoughts and directions in the practice of safety techniques. It is just plain good business for the manager to familiarize himself with the tenets expressed here in addition to the obviously humane implications.

Management Development programs in both industry and academia would do well to incorporate this text and its workings into the most practical aspects of the curriculum.

WILLIAM J. BAKROW, PH.D.
President, St. Ambrose College
Davenport, Iowa

Preface

This book is written to provide company executives with the background necessary to develop safety policies for the efficient, productive, and profitable operation of their firms.

Though this book is a summary for executives, it is intended to be read by all levels of management down to the line supervisor. It can be used as a supplement to current textbooks and lectures in academic courses at business and engineering schools. It can give students an appreciation of the impact of safety on the total business enterprise.

Companies must comply with the statutory requirements of the Occupational Safety and Health Act (OSHA), and the National Institute for Occupational Safety and Health (NIOSH). In addition the Consumer Product Safety Act (CPSA) places a heavy requirement on industry to eliminate safety hazards in their products. The general requirements of OSHA, NIOSH, and CPSA are discussed as they apply to the development of company policy on employee and consumer safety.

Executives require decision making information on all factors that will affect their company's operations, finances, efficiency, and public image. The impact of safety deficiencies that result in adverse publicity, injury and loss of trained personnel, higher insurance costs, repeated plant and product damage, or adverse comment from officials of the OSHA program and other government inspectors, must be brought to light to help executives chart their company's course for more efficient operations. The consequences of improper design of products, resulting in potential hazards to the users, not only will conflict with the legislative requirements of the CPSA but also may result in negligence actions with possible heavy fines, judgments, and public marring of the corporate image. One of the most severe consequences as a result of product safety deficiencies might be the removal of the product from the market, or call-back for modification at the company's expense.

A National Electronic Injury Surveillance System (NEISS) has recently been established by the Bureau of Product Safety of the U.S. Department of Health, Education, and Welfare. NEISS is currently collecting data through a computerized network linking 119 hospital emergency rooms in 30 states to a central computer in Washington, D.C. Subsequent collection of data will be secured from physicians and clinics nationwide. The sum of this material, which will not initially follow from safety deficiencies, will provide information on products *involved* in injuries, although not necessarily *causing* injuries, and will subsequently establish priorities and provide a rational basis for establishing target areas of action by the Consumer Product Safety Commission.

In its drive for corporate efficiency and demand for decision making information, management must not be denied professional internal summaries of all of the impact information brought out above. The objective of this book is to provide yardsticks for gathering just that type of information for decision makers, as well as to provide management with some so-far untapped sources of such vital information.

Some ideas expressed in this book may generate discussion and criticism. Some of them may deviate from axioms and formulas used in current accident prevention approaches. These ideas were intended as yardsticks for measuring the effectiveness of the reader's present safety programs.

There is no single solution to the matter of safety management. It is hoped that this book will serve as a departure point for the reader who must design an effective, economical, and flexible safety program.

April 1975   LEON SCHENKELBACH

Acknowledgments

I sincerely express my appreciation to the following, who reviewed, analyzed, and provided critical and needed comment of the manuscript: William J. Bakrow, president, St. Ambrose College, Davenport, Iowa, and founder of the Motorola Executive Institute; Rabbi Robert M. Benjamin, Temple Emanuel, Davenport; Lillian C. Schwenk, Head of the Safety Education and Research Program, Safety Education Laboratory, Iowa State University, and president, Hawkeye chapter, American Society of Safety Engineers; Harold A. Parmelee, executive vice president, Associated Employers of the Iowa-Illinois Quad-Cities; Robert J. Nixon, president, Davenport Machine and Foundry Company; Robert Kammer, sales engineer, Warner Electric Co.; Alfred Barden, OSHA Administrator, Region II (New York, New Jersey, Puerto Rico, Virgin Islands); Clarence J. Campbell, my former associate and now safety director, Rock Island Arsenal; Jack Silverman, president, J. and P. Consultants; Jerry B. Stout, director of the Muscatine-Scott County

School System; Sister Kathleen Eberdt, Dept. of Economics, Marycrest College, Davenport; L. B. King, plant manager, Caterpillar Tractor Co., Davenport Works; Bess Pierce, secretary of the Quad-Cities Writers Club; and to my many other dear friends in Iowa, who helped and hoped that this book might come about.

In view of the professional and personal status of these people, let me hasten to add that any oversights, omissions, misinterpretations, or errors in the text are my own.

I further wish to acknowledge the Midwest winter, with its cruel, icy blasts which seem forever, and which confine one to home and creation, and provide the opportunity for clear and unconfined thought and endeavor.

L. S.

Contents

1

Background

The basic philosophy of safety management is to eliminate human suffering and anguish, and to achieve economy of operations, in that order. Regrettably, the area of prudent safety management is often considered one of the nonproductive disciplines, and is one of the first to feel the axe during a period of retrenchment. This is a decidedly short-sighted approach. No function of management which has the capacity to save money for the company—sums of money far greater than the effort or cost expended for the saving—should ever be considered nonproductive.

This book is aimed at alerting corporate management to the need for professional management of corporate safety requirements, where such management may not now exist to the desires of the top executives. The complexities of the task demand professionalism. Safety requirements will multiply rather than diminish, and industry must meet them.

Only a small percentage of U.S. companies maintain a full-time professional safety staff (even if that "staff" is only a single manager) to direct their activities in the safety area. Generally, some one individual in an organization is singled out as the "safety director," "safety officer," or "safety engineer" to cover the safety assignment—in addition to his other duties. Such other duties may range from full-time work in industrial or personnel relations to work in the medical department, work as the plant engineer, or work as maintenance foreman, etc. The title "safety director" looks good on an organizational chart; it meets some of the requirements of government to maintain an aggressive safety program as legislated by the Occupational Safety and Health Act (OSHA); it shows an aggressive attitude to the company's insurance underwriters—*but does it accomplish its nominal objectives?*

The individual designated as in charge of safety is frequently harassed and overburdened by the duties of his other assignment, so that the time he spends overseeing safety activities either is nil or minimal. He may not like it that way, but he's bogged down in the details of his other major assignment and can do little about it. Quite often, his only relationship with safety is the title bestowed upon him by management—but this does not mean he doesn't want to learn and show his mettle with respect to safety.

Too often the corporate or plant "safety director" has received little or no formal training in that profession. He may either have been reined-in to safety by management, or not permitted the time or funds to broaden his perspective. His only exposure to safety may have been accompanying insurance representatives or government inspectors on their rare visits to the plant. These visits, of necessity, are often hurried, and the exposure barely scratches the surface of the basic elements of prudent safety management.

Many if not most "traditional" actions in short-stopping the safety functions of a company are basically shortsighted. Let's see why.

Known factors (the tip of the iceberg):

1. Accidents cost money in *compensation* (and *insurance*) costs.
2. Accidents cost money in *personnel losses* of trained (or untrained) employees for a day, a week, or longer.

Hidden factors:

1. Accidents result in *material and property damage* where too often the costs are hidden under the rug in "operational costs" or "maintenance of plant and equipment."
2. Accidents cost money because of *work stoppages* by all concerned, and all who discuss it within the plant. In some cases, a severe accident will bring a plant's activities to a complete halt.
3. Accidents result in delays in filling orders, and may cause unsympathetic feeling on the part of customers.

And so on, ad infinitum.

But this is only the beginning of a tabulation of the costs. There are, however, ways to lay out a basic safety program without unwarranted costs, a program which will not dislocate staff operations, which can be launched almost immediately, which will bring hidden costs out from "under the rug," which will bear fruitful results in conservation of manpower and funds, and which will reduce work stoppages. Such a basic program is not the kind of cure-all extolled by the poster putter-uppers, the handout giver-outers, and the well-wishing individuals who feel that the accomplishment of a total safety program is to give everybody a pair of safety glasses. These items are important within their own sphere, limited as it may be, but they in no way signal the accomplishment of a broad program.

This book will attempt to be concise, while at the same time providing the basis for the establishment, direction, and management of a total corporate safety program. The success of such a program can be measured by some of the yardsticks which will be provided herein, but the program's ultimate worth can only be realized if it is effectively put into application.

The basic data of such a program can be used by the largest companies, or scaled down to the smallest. True, some companies do not have staffs, per se, and middle management may be small indeed, with the top executive being involved not only in policy-making and general planning, but also active in the shop areas. In such cases, as promoter and supervisor, he then assumes a more active role in a safety program. Whatever the size of the company, however, the basic elements brought forth, if practically applied, will not only reduce accidental losses and so reduce the red ink side of the ledger, but above all will instill in employees an increased sense of respect for management—and thus make them more productive.

Some of the views and programs put forth in this book may sound extreme, but it is my belief they are not. They are backed by hard and sound fact, and by the view that too often in the interest of expediency the short course is taken. How often has a "calculated risk" really been calculated? From personal dealings and related experience with a considerable number of companies, I have become firmly convinced that too often we throw away dollars to save pennies, and that the real risks have never seriously been measured.

2

Top Management— the Chief Executive

Top management determines the objectives of a company, and outlines the policies which must be followed to reach those objectives. In order to effectively create policy, the chief executive must depend on information generated by his subordinate managers, and part of this information must come from those with the responsibility for safety involvement. To eliminate the input of the safety manager is to dilute the necessary team effort and to deny top management access to precise decision-making facts. We'll see why.

The basic policies established by the chief executive of a company, and the efficiency with which such policies are implemented, will determine whether or not the operation of the organization will be profitable, and —consequently—the degree of success of the executive. Responsibility may be delegated, depending upon the structure of the organization, but ultimate accountability for success rests in the hands of the top manager. This applies not only to product development, manufacture, sales, personnel policies, and maintenance of plant and equipment, but also to the cost accounting that accompanies the accomplishment of these functions.

One of the basic elements of a company's total activities in the achievement of economic operation is the capacity to minimize or eliminate waste created by accidental loss of manpower, material, or production capacity. A successful loss prevention program will reduce the factors of human loss and economic waste, and will thus raise the profit margin. To achieve this, safety controls should not—as they often do—stand alone, but rather should be woven into the entire pattern of company management and operations. A "minor" accident may go unnoticed in the everyday work activities of a company. A major accident, with the capacity to wipe out a plant or a department, or with such severe effect as to undermine the morale of the work force, will result in disastrous economic impact. But a series of minor accidents can be as disastrous to a company as one major accident—and if management is not aware of what is happening, no one will ever know.

It should be understood that a safety program is not a one-shot effort, but a continuing process. The chief executive should direct that safety policies be incorporated into all directives affecting every phase of his company's operations, and require periodic reports as to how the policies he has outlined have been implemented. The attitude of top management will then be reflected throughout the organization and plant, once its policies are supported by middle management and reflected by line supervision at the point of production.

Without a sincere and concise statement of policy by the boss, no safety program can exist. It will flounder at some intermediate level and either may or may not develop into a paper or "eyewash" program, if it continues to exist as a program at all. Net result: repetition of losses, faulty operational practices or procedures swept under the rug, undermined morale, defective product, and dissatisfied customers.

ACTION

1. Policy Directive(s).
2. Presentation at highest staff levels.
3. Statement that such policy be reflected in functional activities and internal procedures.
4. Designation of responsibilities for implementation of policy.
5. Demand for periodic reporting on actions taken to implement policy.

One effective approach and statement of such a policy was made by a Commander of the Military Ocean Terminal (a port within a port) in Brooklyn, N.Y. On assuming command, and before his initial staff safety review, the Commander requested a concise safety audit of the organization. When that was given him, he reviewed it carefully and jotted some notes. At the review, the Commander's words to the staff (military and civilian) went something like this:

> As Commander of this installation, I bear the final accountability for its mission. As you know, the responsibility for effective operation of the various divisions has been delegated to you, and the ultimate success of this command rests with you and all of the personnel under your direction.
>
> The final accountability for the operation of this com-

mand's safety program, however, rests with me, in this office. It is not that of the safety director, or of any of you other managers, but it is mine.

As my personal staff representative in that area, the safety director will coordinate with me and with you in the direction of the total program. The responsibility for the safety program's successful implementation within your areas of responsibility rests with you. I will desire periodic reports of all phases of that program, its accomplishments, its pitfalls, its results.

We all fully realize the possible catastrophic effects of a major accident on the waterfront or, for that matter, anywhere else in this organization. We also fully realize that accidents of so-called lesser impact have more serious potential. The success of our safety efforts, therefore, will be reflected in how well they are implemented by every one of us in every job in this command, to prevent the creation of circumstances which would breed accidents of any degree.

The safety policy of this command is to eliminate potentials for accidents wherever they exist, and to reduce them to the irreducible minimum, within our total capability and resources.

None of us may reasonably assume that any operation is a good operation if it is not performed safely. No operation may be considered efficient if we sacrifice the safety of our personnel or physical resources, or only provide it lip service, in the interests of expediency. To this end, I desire that:

1. Every manager direct the safety program within his own area of responsibility just as if he were in charge, and bore the final accountability.
2. All other managers and supervisors reporting to you assume their similar responsibilities within their own departments or work areas.
3. All employees be advised of their personal responsibilities for safe work performance.
4. Safe performance be a criterion for promotion for everyone in this command.

I would like this policy to be reflected in both written and verbal communications within your own divisions and departments, and brought out in discussions between supervisors and employees, so that it is clearly reflected in every nook and corner of this installation. This will not be a one-shot affair, but an ongoing thing.

I intend to make periodic visits to all departments and every area of this installation, not only for personal, operational, or administrative purposes, but also to specifically observe how well this policy is reflected in the working habits of all employees.

This policy statement of the Commander to his staff was like a shot of adrenalin, and it had its desired effect. Every manager knew where he stood, what to do, and what standards would be expected. In addition to the official and unofficial communications and discussions that followed, the grapevine spread the word rapidly, and everyone knew just where the "old man" stood. The statement worked, and worked well. With the cooperation of the stevedores and longshoremen, the following came about:

1. Disabling injuries were drastically reduced.
2. First aid cases were similarly reduced.
3. Property damage costs took a nosedive.
4. Cargo damage costs took a nosedive.
5. Total audited accident costs were reduced considerably.
6. The longshore accident rate took a plunge, and reached an enviable low level possibly never before experienced by any other similar waterfront operation.

In large part because of the Commander's stand on safety, the smooth movement of people, cargo, and ships proceeded efficiently, on time, and without the injurious and hindering dislocations and delays which might otherwise have been created.

One of the significant things about the Commander's presentation was that it was expressed in his own words and thoughts. It was genuine, and the actions taken throughout the organization reflected a response to his sincerity. Had his policy statement been a ghost-written presentation, it would have been recognized as such, and lack of action by subordinate management would have mirrored that recognition.

Management is always interested in safety, but the degree of its interest which is manifested in action will determine the success or failure of a safety program. Since managers are responsible, for (among other things) persuading people to work in accordance with required standards, the responsibility for safety performance must accordingly rest with management. Effective exercise of this responsibility necessarily focuses on the uppermost echelon of management, the top executives. Top executives must demand:

Investigation—to identify risks.

Evaluation—of risks.

Elimination—of risks that are intolerable.

Assurance—that tolerable risks are controlled to prevent severe accidents.

Correction—of uncontrolled risks.

Follow-up—periodically, to assure maintenance of controls and to ascertain that no new intolerable risks are introduced.

Reporting—periodically, to determine effectiveness of action by lower echelons of management and supervision, and to be informed of the complete accident loss posture of the organization, this to include:

Personnel Injuries (lost time and first aid).

Production delays due to accidents.

Property damage due to accidents.

Product damage due to accidents.

Shipment delays due to accidents.

Scrap loss due to accidents.

Rework cost due to accidents.

Maintenance costs due to accidents.

Design changes to eliminate accidents, injury, or adverse effects on health.

Engineering changes to eliminate accidents.

Operational changes to eliminate accidents.

In-depth accident investigations to determine the root cause(s), rather than the surface or immediate causes.

Elimination of the word "careless" in accident reporting. (The word "careless" has been stretched to encompass everything, say nothing, and be utterly meaningless. It fails to pinpoint the how, what, and why of accidents.)

Periodic reports from individual departments reflecting the financial impact of accident losses to the company. These may be reflected in the normal financial reports which gauge the effectiveness of departmental operations.

None of this information-gathering requires a massive expenditure of funds. Such information-gathering may appear extensive, and possibly overdetailed, but its purpose is to bring basic management functions to the surface.

A department's accident profile often reflects its operation and production capability. When the frequency and severity of accidents begin to rise, or when accidents begin to snowball in a "quiet" area, an analytical review can uncover drastic management deficiencies and possibly other operational malpractices reflected in undue costs. A major benefit of this long-range total program is to provide a sounding to gauge the effectiveness of management's ability to manage.

Top management must emphasize that one of the criteria for promotion is safe performance . . . whatever the individual's level in the corporate hierarchy.

3

The Staff

Staff supports top management in meeting and solving the problems of an organization. The basic functions of staff are to:

Advise and provide information.

Develop plans.

Achieve coordination.

Make decisions.

Prepare and transmit directives.

Maintain control.

Top management prescribes the policies and establishes the goals. Staff translates them into action.

Once having received from top management the basic policy guidelines concerning the safety posture of the organization, staff must react decisively. Staff cannot have "tunnel vision." The definition of the term "tunnel vision" sets forth by opposites what staff must strive for: to coordinate with other staff elements, and assist each other so as to take advantage of all of the

14

combined know-how; to pick each other's brains and agree on a course of action; to resolve all problems among staff, and present the finished plan (if one is in order) as finished, and not as tentative. Top management depends on this type of action by staff and expects programs finished in accordance with its policies.

After such combined staff action, each staff representative must then provide the guidelines and direction for safe accomplishment of his element of the program within his own area of concern. This may be done through meetings, directives, instructions (verbal or written), or suitable operating procedures to the next lower level of management or supervision.

In such a sequence, safety must not be considered a separate item standing apart from the other factors in the overall plan; rather, it must be woven into the fabric of the plan to incorporate all known or anticipated hazard control procedures. Time and money enter into all of your plans, but a major point of consideration is "How much can you afford to lose?" The answer to this question is of vital concern to top management.

In determining the answer, it is imperative first to accurately determine how much is being lost through safety deficiencies, right now. Some of the measurements of effectiveness are brought out in Chapter 9, and costs are discussed in Chapter 10. The answer to this question, in conjunction with the chief executive's broad reporting requirements discussed earlier, and an analysis of the company's product safety posture, will provide a clearer perspective of the company's loss position. In reviewing the following individual action areas of various staff managers, each representative may then accurately determine whether possible lack of research or the broad desire to save money is actually providing an overall savings to the organization as far as loss prevention is concerned.

Product Design staff must consider how to design the end item so that it is not only a better product than competitors', but is tolerably hazard-free and will not result in subsequent legal actions by injured customers. During the design process, in-depth reviews should be made to design-out any problem areas which may not only cause a future accident (within reasonable limits), but may result in the end product being returned for rebuild or rework. Product design should also be reflected in the elimination of production, packaging, and shipping hazards. An in-depth review of this area is fully explored in Chapter 11, "Product/Systems Safety."

Engineering and Industrial Design staff, in addition to supporting those in product design, must provide for a smoother work flow and safeguard the working environment to prevent accidental loss in the steps from initial fabrication through final shipment. In their coordination, these staff elements might well consider a systems approach whereby safe design is incorporated into the base product and is maintained through all process operations from receipt of material to final packaging and shipping. This will prevent good design from being ruined on the production line or on the shipping platform.

Personnel (or Industrial Relations) staff must take a closer look at new hires, and consider physical examinations for personnel involved in high hazard or strenuous operations, so that the company doesn't "buy" a potential liability or compensation case at the outset. If many employee defects or medical deficiencies are not determined at the time of hiring, who is to prove that these did not later result from occupational hazards while the person was in your employ? Personnel staff must "sell" the new employee on the safe working environment and the shipshape condition of the plant.

Let's face it. With our current technological advancement and shortage of trained labor in many areas of this country, the stable labor pool is often limited. Labor will no longer come to work in a plant that's a "dog" when other opportunities are available. If your plant falls into the "dog" category, the grapevine will put the word around and you will then become the employment point of last resort. You can't hide it, in either a large or a small community.

Operations (or Manufacturing) staff must find out the most economical way to produce the product, with as little time wasted as possible. It must take a close look at prevention of "downtime," production delay, and waste of manpower and materials. A serious accident (or one even thought to be serious) has the potential of stopping an assembly line for a day, and silencing a whole department. The operations staff element must assist in planning for and insuring an integrated and uninterrupted work flow. All known or anticipated problems must be incorporated into the basic work plan. Some anticipated problems include bottlenecks in process operations, poor supervision, slipshod industrial housekeeping, and congestion. The latter factors will skyrocket costs, since the shortest travel distance between two points of operation may become a twisting path.

Purchasing staff should re-examine its approach to procurement of machinery, raw materials, and equipment, bearing in mind OSHA requirements. It's more economical in the long run to purchase equipment which is protected within existing standards than to modify it locally. If inert chemicals and related materials can do the job, why not consider their purchase rather than high-hazard, flammable, explosive, toxic, or corrosive items? The latter only create problems for Operations, Engineering, and Maintenance in providing safeguards for their use and storage—increasing the frown on the comptroller's brow because of excess expenditures for such protection and the potential for higher insurance costs. Here, as elsewhere, it is clear that the expenditure of a few pennies can save many dollars. It is further imperative that purchasing people consult with members of staff prior to procuring any item which may be potentially hazardous. Such consultation may take only a moment but may ultimately prevent serious injury, with resultant monetary loss to the organization. *This must be standard policy.* To do otherwise would be to save a few pennies initially but court a problem later.

Maintenance staff must plan for the smooth working of the plant, machinery, and utilities, and must insure against stoppage of those facilities through breakdown. The success of a company's maintenance staff is directly reflected in production. A typical example: the frequent blowing of air-receivers will eliminate, and thus prevent, discharge of sediments through the air systems and their eventual deposit on finished products by way of the paint spray operation.

Too often, Maintenance becomes the whipping boy for other departments. Besides taking care of fair wear and tear, and keeping the plant humming, maintenance people must also repair the damage caused by faulty procedures within the plant, and by uncontrolled moving equipment, fork-lifts, etc. Rather than pointing the finger of blame at Maintenance for not fixing everything yesterday, we might inquire as to the reasons for the damage in the first place. By so doing, we can then logically charge the department at fault with the repair costs, rather than assigning it to Maintenance, which has a difficult time meeting its own budget without the addition of nonincurred costs.

Let's look at the comptroller, or his small-company equivalent. The comptroller is the person who must take a close financial look at every segment of the company and make an analysis of what's spent, and why. He must ask the question, "Does the need justify the cost?" If he merely looks at surface costs he has not done his job. He must look beneath surface costs to the reasons for them, to the determination of whether or not the end result warrants the expenditure, particularly when the budget has been stretched thin. Some aspects of this are elementary. The determined guardian of the purse strings has the capability of digging deep into the *why*, the *need* for the expenditure.

The comptroller, in keeping a close watch on the purse strings as well as the efficient operation of the organization, inherently dislikes waste, or the potential for waste. In attempting to reduce losses, it would be well for him to be keenly aware of the hidden costs (or losses) which are pointed out in Chapter 10. It's amazing, for example, how costs that are glaringly obvious to a safety professional may be hidden from the eyes of others. We could go on and on through all other staff elements, but why not have all of them draw up their own basic plans and requirements? They're specialists.

4

Middle Management

The middle manager is the wheelhorse. He receives policy and program from top management (or staff) and must translate this into action, not only within the radius of his immediate boss's views, but also within the directives of the chief executive. He is directly responsible to the former, but should never forget that the latter is running the organization. Middle management, then, translating plans into action, backs up top management and staff by attempting to find ways and means to produce the product better, *with conservation of men, material, and money.* The middle manager must make every effort to eliminate or reduce waste of human and physical resources.

In establishing the working guidelines for line supervision, the middle manager:

1. Insures the implementation of top management's safety policies by making known and supervising the executive's desires.
2. Acts as the boss's field manager by providing area guidance to carry out the stated requirements.

21

3. Maintains broad supervision by supervising the supervisors, and insuring that they support executive safety policy.

4. Provides personal attention to and participation in production, and may doubly prevent disruption through accident.

5. Keeps personnel informed and motivated. The result of this is high production, fewer errors. When important needs are unduly thwarted, production and morale are low, and an accident susceptible environment is induced.

6. Realizes that intense fear, anxiety, stress, and fatigue induce "industrial casualties."

7. Devises new methods and prescribes working directives to achieve top management's standards, always with an eye to providing a safe working environment.

Since his immediate operating budget is always under close scrutiny, and must cover losses due to delayed production as well as the remaining factors covered in Chapter 10, he must manage his operation through precise planning, organizing, direction, and control—in other words, the basics of management known to successful executives.

In clearly making his views on hazard control known to line supervision, and following up to see that his objectives are carried out, he must insure that:

1. Line supervision is fully aware of its safety and total loss prevention responsibilities.

2. This awareness and action response is reflected in the work force.

3. His "house is shipshape" for smooth product flow, with an unencumbered work environment.

This manager, being in the arena of action, may determine how mechanical means, gravity or otherwise, can eliminate the need for brute physical force, and prevent injury to trained workers, as well as damage to raw material or product. Having determined this, he may then convey his justification for change to his boss. (If everyone had done their homework, though, the staff elements in their basic planning and coordination with their middle managers might properly have resolved the matter in its planning stage before it became a problem.)

5

The Line Supervisor

The line supervisor is the management representative at the scene of action. He's responsible for completion of the end product and is in closest association with the working environment. He knows the individual employee better than anyone else in management. It is up to him to see that the basic policies of top management and all intermediate levels are carried out. It is he who must determine whether the individual employee is adequately trained and capable of carrying out his assignment, even if this assignment is merely pressing a button or moving a switch. It is he who knows whether that employee is fit for the operation on the day and at the time that it must be performed.

In producing the end product, the line supervisor must know whether or not the individual can accomplish the job on a given day, and whether or not that individual is in the frame of mind to do the work. Any employee, on any particular day, for reasons known only to himself or herself, may be incapable of properly

24

performing what's required or expected. And the quality of the end product not only depends on the smooth working of machinery and flow of material but also on the nature of the individual. The line supervisor, knowing the individual characteristics of his own personnel, may know where to shift people, where to assist, and where to back off. He must!

Since he supervises employees' everyday activities, he must be the one to enforce (not necessarily in the dogmatic sense of the word) the proper procedure for doing the job. If he sees something being done wrong (whether it be an improper operational procedure or an unsafe action) and says nothing, it must therefore be assumed by the employee that the action is proper. If he corrects an incorrect action, then the employee is capable of accepting the fact that an error has been made. But silence in the face of an incorrect action tacitly condones that action, and can only result in repeated incorrect actions, the end result being ruined work, procedures disrupted, injury, and property damage.

Now, since the line supervisor is at the scene of action, should he not have the responsibility of seeing to it that the job is not only correctly but also *safely* performed? As a matter of fact, no operation which is performed unsafely is correct! Should not the supervisor be concerned with the welfare of all employees?

In supporting his boss's and top management's principles and guidelines, and in adhering to company policy, the line supervisor must realize that he is on a rung of the corporate ladder. The line supervisor is many things to many people, but he is—first and foremost—management's direct link with the person in the shop. He must also be chaplain, timekeeper, mediator, diplomat, and shirtsleeve psychologist.

The line supervisor is also management's basic contact in another vital area—safe performance. In the course of his everyday activities, does he take time to sell safety to his people? If not, then we have not convinced him of the necessity of working at safety in conjunction with production. Safety, to a great extent, is an attitude, and we must insure that this is exhibited by the line supervisor so that it is then reflected by the employees. Without this, any safety program will ultimately flounder, with resultant adverse effect on production.

6

The Employee

Employees are the basis of production, profitable or otherwise. Management seeks good workers, and strives to keep them. In many areas, the labor pool of trained, stable people is dwindling. Coupled with this is the fact that a well-trained employee in a good frame of mind will out-produce, and out-perform any run-of-the-mill person, even though that person can be immediately put on the payroll.

With the shifting pattern of growth and expansion of industry, and with a more mobile U.S. population than ever before, the employee can always look for a better place to work if he is dissatisfied where he is. The automobile has given him that mobility and he uses it to his best advantage. He will naturally prefer to drive a few extra miles to work if it is to a more advantageous work environment, and he will not accept autocratic or dogmatic management if he can do anything about it.

Industry must accept these facts and make provision for retention of a trained labor force. This does not mean that the employee must be or wants to be mollycoddled. He seeks fair, affirmative, and determined leadership, and wants to feel that he is more than a numbered timecard in the corporate or small business world. With proper treatment he will stay and be productive, relieving the strain and drain on Personnel's (and Supervision's) capabilities.

No employee, no human being, however, can achieve and maintain full productivity in an environment where injury, no matter how minor, takes a daily toll. No employee can be effective if he feels that he may be the next one injured, the next statistic on the compensation rolls, with possibly little sympathy from management. As a human being, what is his reaction when he senses the immediate possibility or potential for injury on the job? What would your reaction be? With proper supervision and management, however, and with the proper safety attitude displayed by his boss, he will generally cooperate and go all-out to assist, and will be the first one to correct an improper or unsafe action by one of his coworkers.

But of course the employee must do his part, as well. He is the person producing the end product. He is the person at the machine, or assembly line, or moving the goods, or stacking the material, and it is he who must do it right to keep from ruining the end product or injuring himself. One good barometer of the effect of production (and supervision) capability, besides the accident rate in the department, is the scrap and rework ledger. Nothing is more frustrating to management than to finally produce a good product, and then have it ruined by haphazard stacking or handling, or shipping on the loading dock. A slipshod working environment is a bad working environment, and it is people—management and employees—who control the working environment. We may not be able to control the configuration or age of plant and structure, but we can control the methods of production which will make the best of the facility, materials, and people with whom we must work, and who eventually determine the color of ink on the final balance sheet.

Employee attitudes may be generated in many ways. Sometimes it takes the compassion and fire of a Knute Rockne, sometimes it takes a cold, dispassionate academic approach, and sometimes it requires individual acts of the imagination to reach people, always depending on specific circumstances and conditions. The supervisor or manager who is in close touch with the employees is most apt to sense the approach that should be taken. Here, for example, is an unorthodox approach that the author took when tied to a very limited budget. We had admired the impact value of those funny candid baby pictures with illustrative captions. Rather than publish the expensive copyrighted photos (which we couldn't afford anyhow) and insert our own captions to fit our specific problem areas, we ran a little contest. We asked employees to submit photos of children in their families, preferably in spontaneously contagious and comical poses. The selected photos would be enlarged, with a suitable caption giving credit to the submitter. The results far exceeded our expectations.

For example, a photo of a child with head cocked, wide-eyed look, and hands on head.

Caption: "_____, daughter of _____, welder in machine shop B, says, 'Don't use that broken equipment. Wanna get clobbered? Get it fixed!' "

For another example, a photo of a child with wistful, chastising expression.

Caption: "_____, grandson of _____,

hatch foreman, Pier A, says, 'If you tell me to keep out of the street, why do you walk in that equipment lane?'"

And so on. Sure, to some it may seem corny, but in this instance it worked, in this intensely people-oriented world, and successfully supplemented all other safety publicity approaches that we took. It boosted morale, brought our own people into the act, and the photos with captions put across the message. They were not a cure-all by any means, but they were some of the most effective material that we had to supplement our overall total program, and they did help to create awareness, motivation, and morale — all of which were reflected in *attitude*.

Many years ago, the author, upon accepting a position as safety director on the Brooklyn waterfront, had to start learning all over again. Having been involved in several aspects of safety, and having previously served as a safety manager in an organization of massive scale, it didn't take long to realize that here was a totally strange and new environment. To be effective, the ways of the waterfront had to be learned, and the intricate techniques and procedures of that most hazardous occupation, longshoring. At my request, the stevedore (the boss) set up an initial meeting with his hatch foremen (the supervisors) and his key managers. After a brief discussion of what my homework had revealed, and admitting to them that I was a novice to their business, I requested that they train me. (They had an excellent grapevine, and already knew my entire background — what I could do, and what I couldn't do.) I asked that they let me know in no uncertain terms everything that I did wrong in my rounds of the piers and ships, as well as the various activities surrounding them. They agreed, and said they would pass the word around.

Try as I might to do right, I blundered miserably, over and over again, and with each stupid mistake that I made, one or more of the longshoremen (the workers) would bellow at me. On occasion, a brawny but friendly arm would pull me out of the danger zone, and shake up my eye-teeth in the process. They told me what I did wrong, how it could get me clobbered, and how to do it right. It didn't come easily, and it didn't come fast. It took time—but it paid off. Together with our own Army personnel in and around that waterfront, they drummed some savvy into my head. I learned the bits and pieces, the jargon and names of the many things that encompass that world. Above all, once again, I learned how little I knew.

The longshoremen appreciated my original comment about my lack of know-how. Had I kept silent and attempted to bluff my way through, they would have known it in five minutes (or less), and I would have lost my credibility. I respected their advice, had the best of instructors, and can never thank them enough for it. Through personal involvement I gained a great respect for those people and the hazardous working environment surrounding their activities. Where I would have bucked a stone wall had I bluffed, I learned to relate to them and their working environment, and got their cooperation.

One major side effect of all this, and perhaps the most important result: By correcting me continuously, with bellowing, action, and down-to-earth advice, they bolstered an acute awareness within themselves, bosses, supervisors, and workers alike, of the need to do the right thing so that they wouldn't be placed in the position of saying "Don't do what I do, do what I tell you to do." In short: management, training, attitude, environment, teamwork, control. The employee seeks this, appreciates this, and will be more effective because of it.

7

The Plant

A plant is a combination of people and machines, teamed up to produce a certain line of products. It may not have been originally designed for its current operation. It may be old and outmoded and have changed its character many times over, but it still should be functionally capable, within its limitations, of producing the present end products in the most efficient manner and with the greatest economy possible under the circumstances.

Coupled with the physical design of plant and structure, is the approach taken by management to reasonably tailor the operations to suit the facility. If operations have been properly planned and the factory efficiently maintained and suitably managed, the plant can more probably show a profit. Perhaps the plant was laid out some time ago for operations which have since been changed. If the nature of the work flow has not been modified to suit the manufacture of current production, but still maintains the same layout, procedures, and characteristics of fabrication of some item

33

long ago lost in the history of the establishment, then the current method of operation is probably inefficient. We are attempting to produce modern items, to meet modern competition, with outmoded production systems.

The fact that the plant may be old does not necessarily make it a bad plant, however, if thought has been given to keeping up with the trends of modern manufacture. Responsible management knows this axiom; That along with the basic laws of supply and demand, profit depends on production of the end item at least cost. Least cost does not necessarily mean letting a plant run down and forgetting everything but immediate "necessary costs." Too often, businesses fail and companies either disappear or get swallowed up by more efficient enterprises simply because the product is pushed out without sufficient thought to modification of procedures to keep up with the times. Then comes the sudden realization that the world has changed and that the old manufacturing procedures simply can't cope with modern competition.

If you must use an outmoded structure, for economic reasons alone, it would be wise to consider that fact in its complete perspective. There is nothing that you can do about it from the immediate structural point of view if the budget is tight, but you can thoroughly review the basic layout—from receiving area, through the fabrication and packaging processes, to the shipping point. You must capture space. Work-flow processes must be tailored. The work force must be retained. Considering your limitations, the work environment must be capable of avoiding built-in obsolescence, and of meeting the challenge of production and personnel retention.

Let's consider some broad areas. Basic to the work flow is a smooth, ship-shape working environment. Since space may be at a premium (it seems that it always is), a review of the storage and warehousing areas may recapture valuable real estate. Racks and palleting may permit doubling the capacity of the same square footage, providing the floor can stand the load. Controlled industrial housekeeping may open hidden areas. Planned aisle markings and storage points, suitably marked, will simplify and monitor the movement of people and materials, and eliminate going over and around things. The planned layout then results in logical work flow and less backtracking. The combination of all results in less cleanup and straightening-out procedures — all of which cost money.

Add to this some means of controlling machine chips and oil splashes, thus freeing machine operators to do their primary job without taking time to clear their mess for the next shift or the next work day. Plastic or metal shields, and people control, can do the job. The shields are inexpensive and will pay their way in saved machinist's pay and "downtime." You'll also be taking fewer metal splinters out of people, and you'll stop wasting coolant.

Lighting must be adequate for maximum production and fewer rejects because of worker fatigue. Ventilation must be adequate for the same reasons, and for the elimination of health hazards attributable to modern materials and processes. All of the above production assist items are requirements of OSHA, although OSHA is concerned only with the safety factors. The National Institute of Occupational Safety and Health (NIOSH), created with OSHA, is charged with the preparation and implementation of regulatory requirements covering the broad area of occupational environment conditions, and the impact of adequate ventilation in industrial operations bears a high priority. NIOSH has only begun to function, and its demands will certainly multiply.

Inert materials, for instance, should be seriously considered to replace many of the presently used flammables, caustics, and toxic compounds. The cost may be the same, or a bit more or less, but the cost of protection and insurance rates based on risk potential of the old materials is far more. Management would not tolerate, under any circumstances, the introduction and storage of a quantity of raw explosives in an industrial operation or within the working area of a plant (unless it was making explosives and the production environment was designed for it). It follows that that same management should seriously consider the withdrawal of intensely volatile, toxic, or flammable materials from that same working environment.

Gravity and mechanical conveyors and modern material handling equipment will move material fast and efficiently, and with diminished worker fatigue and accident, and will pay their way in short order. It stands to reason that such mechanical equipment should be protected within existing standards (OSHA, American National Standards Institute [ANSI], National Safety Council, Trade Association, etc.) and should be suitably maintained so that it, in turn, does not create hazards. As far as operator-controlled material handling equipment (fork-lifts, etc.) is concerned, intensive training of machine operators should be established to prevent injury to personnel or damage to plant and product. The physical protection of machinery and the control and expertise of operators go hand in hand.

A few dollars' worth of "Exit" signs will show the way for emergency evacuation, and meet some fire inspection and OSHA requirements. Control of smoking in areas where flammables must be used will minimize the chance of fire or explosion. You can be sure that your insurance underwriters did not overlook these conditions when your insurance rates were determined.

It's better by far to secure compressed gas cylinders with chain, or other adequate means, than to have one fall and shear a valve, thus becoming a missile within the work area, or an uncontrolled bomb if it contains flammable gas or oxygen. Nor do plating and stripping rooms need to convey the impression of a complete mishmash, with only a few employees having a clear idea of what is in which vat, or container, or carboy. At the very least with a mishmash, the finished product may accidentally be dipped into the wrong solution, ruining a lot of work—at great expense.

How about pouring the wrong commodity in the wrong vat or tank? We identify hot and cold water taps, and voltage sources of electricity. Why not identify vats and other containers of acid, caustic, chromate, phosphate, cyanide, etc.? It only takes a simple color code striped on the vats. This, along with the stenciling of content and temperature, may prevent many financial losses. In addition, if a worker is injured because of a splash, the cause will be known and the medical and first aid treatment will be simplified.

Two very effective measures were used by two major organizations. In the first, it was corporate policy that all containers, tanks, vats, etc., be identified in a standard manner. Plastic plates of different colors (indicating acid, caustic, water, etc.) were indented with not only the name of the material, but the operating temperature. These plates were affixed to the vats, etc., and were changed as the contents changed. The color code was further applied to the piping and drain lines to and from the container. It is a most effective system, and supplemented the other outstanding safety controls and related activities of that corporation. Besides keeping injuries to a minimum, the relatively inexpensive and long-lasting identification measures greatly reduced product damage in those areas.

The second corporation dealt with heavy nonferrous metals. The product of one major plant involved massive extruded, machined, and rolled products. Of necessity, these same items had to be lifted by overhead cranes and deposited in various soaking vats for the finishing process. A system of color coding and identification was expanded to include the tops of the many tanks, so that the overhead crane operators could identify their contents. Additionally, the tops of other significant items in the plant were painted with cautionary color identification to prevent overhead damage. Before the system was installed, product and plant damages had been quite high. After the total identification process was put into effect, such losses plummeted. A side effect was an increase in production, since the brightly outlined and marked working/storage/tank zones provided perfect targets for the overhead crane operators.

It is illogical to maintain on-site stores of huge quantities of dangerous materials, such as flammables, toxics, cyanides, caustics, and other chemicals, when only small quantities are used daily at points of operation. Why not store them elsewhere, in a controlled environment, and use the released floor space to better advantage, with less potential for serious accident/injury/plant disruption? If exterior storage is available, it might be wise to use it to get dangerous materials out of the plant—and at the same time realize a good reduction on your insurance rates.

Mixtures of incompatible materials may result in fire, explosion, or generation of deadly gases. Too often, in reading a news report of the result of a disastrous fire, one notes that "suddenly the fire was punctuated by a series of unexplained explosions." Unexplained to whom? Certainly not to the trained manager or specialist who fully knows the dangers of some of the materials used in modern industry. Knowing this, why not remove these hazards to the best of our ability, or replace them with inert materials that will do the job?

Perhaps the housekeeping in our paint-spray areas or flammable storage points may be improved considerably, to prevent the accumulation of harmful vapors in the flammable or explosive range, or prevent spontaneous cumbustion. No costs—just forethought, effort, and know-how. . . . We could go on and on, but the key point here is to stress planning and control, together with training.

The previous hazard areas have been discussed at some length. The reasons for this are that these serious problem areas are quite common to most industries in varying degrees, are major breeding grounds for injury and damage, and are heavily reflected in insurance costs. Through a concentrated effort in correcting deficiencies which may exist in these situations, the company will make broad inroads not only in the elimination of many of these violent or insidious hazards, but also in reduction of costs — as well as compliance with many OSHA and other mandatory and advisory standards. At this time it would be in order to discuss in brief some of the principles of the hazards concerning flammable, toxic, and other dangerous materials; some broad areas for control; specific situations in representative plants or industries; and policy guidelines to provide management with the tools to trigger action for such control.

FLAMMABLE/HAZARDOUS MATERIALS AND CONTROLS

To enumerate all the flammable, chemical, dangerous, explosive, and toxic materials and their control requirements would require many volumes. Appendix A lists OSHA sources and other organizations originating standards. Appendix B lists various commercial and other reference sources from which technical data concerning these materials may be obtained. As an example, the 1974 edition of the NIOSH Toxic Substances List, prepared in compliance with the requirements of the OSHA, contains 42,000 listings of chemical toxic substances; 13,000 are names of different chemicals with qualifying toxic dose information; and 29,000 consist of synonymous names and codes which have been acquired from published literature, cooperating industries, and the American Chemical Society. This revision includes approximately 5,000 new chemical compounds which have not appeared in the 1973 edition.

Some of the most commonly used materials in industry today are solvents, many of which, like naphtha, toluene, and methyl-ethyl-keytone (MEK) are toxic, flammable, and explosive, yet are still frequently used for dry cleaning and degreasing; followed by paints, thinners, varnishes, and turpentines. Other materials include chemicals; acids which produce corrosive vapors, toxicity, and burns; powered metals (e.g. beryllium, zirconium, magnesium) which can be explosive when in fine dust suspension; gases, e.g. LPG, acetylene, and other flammable/explosive types; oxygen, which, though not flammable in itself, violently supports combustion and should be treated as a flammable gas; ammonia and chlorine, which are both toxic, etc.; plastics of various types used not only in the plastics industry itself but as parts in industrial fabrication, and which may be flammable or give off toxic gases when exposed to flame.

Common examples of volatile flammable liquids are ether, acetone, gasoline, ethyl alcohol, methyl alcohol (wood alcohol), benzene, and toluene. These liquids are quite volatile and at room temperature, if unconfined, may evolve vapor concentrations in air within the explosive range. Some flammable liquids such as linseed oil, paints, varnishes, and enamels may, under certain conditions, be subject to spontaneous ignition and must be kept where any heat produced will readily dissipate. Only noncombustible sweeping materials should be used for cleaning up fluids of this type, to prevent creation of further hazards.

SOME BROAD AREAS FOR CONTROL

Flammable Liquids. As a general rule, flammable liquids do not themselves burn. It is their vapors given off during evaporation which ignite and burn when combined with oxygen. Most flammable liquids emit vapors heavier than air which will seek the lowest levels in an establishment or household and settle in depressions, pits, on floors, and in basements. If the vapor concentration is in the flammable or explosive range, any source of ignition, whether it is a match, cigarette, mechanical or electrical spark, or static electricity, may ignite the vapor which may then flash back to the source from which it emanated, its storage tank or pail, and ignite that source.

The structure of the fire spectrum consists of three major factors: fuel, oxygen, and a source of ignition. A major effort should therefore be made to control these factors, primarily to eliminate one of the elements (the source of ignition) and therewith the potential for fire. It is not always easy or possible, in industry, to separate and eliminate one of these elements, since fire, in one form or another, must be used in various industrial operations. My first priority would be to substitute inert materials for the dangerous ones, if at all possible. If this cannot be done the next step should be to minimize the quantity of combustibles in any one operating area so as to reduce the magnitude of possible fire. For this reason, policy should state that minimum quantities of these materials are to be used at specific points of operation, within operational considerations, with the mass stored in an isolated, protected area. When used in industrial (and other) operations, flammables should be kept in closed, approved containers (this is a long-existing National Fire Protection Association [NFPA] standard, and an OSHA requirement) in order to prevent escape of vapors with their resultant potentially hazardous effect (the fuel). When these flammable liquids are handled to any degree they are usually exposed to air at some point, as when filling or emptying containers, mixing, or transferring liquids from one container to another. Vapors are released during these activities, creating potentials for fire, explosion, or health hazards. In order to keep flammable vapor concentrations to a minimum, the exposed flammables should not be handled in quantity, where practical and possible, and should be kept in approved, closed containers. Leaky or damaged containers should be discarded without hesitation.

Control measures should insure that all possible sources of ignition are kept away from both the flammable liquids themselves and their vapors. Smoking should be rigidly controlled within a satisfactory distance established by fire authorities or other specialists. To simply post a "No Smoking" sign is often ineffective, because the distance or area in which there should be no smoking is not properly described. (Ask any three people in an industrial plant how close they may smoke to a "No Smoking" sign and you'd probably get three answers.) Spark- and flame-producing devices should be prohibited, and the electrical equipment and fixtures in that area should meet existing standards.

Adequate exhaust ventilation, not only at high but at low levels, should be provided in all areas where flammable liquids are used, handled, or stored, in order to dissipate and remove the vapors, as well as to prevent the buildup of a flammable/explosive/toxic range.

Chemicals. Some chemicals or fine metal powders which seem to be harmless by themselves may react vigorously upon contact with commonplace substances such as water, flammable liquids or solids, and the like. The result may be fire, explosion, or toxic gases. For example, metallic sodium may be stored in airtight steel drums, yet in combination with water it reacts violently, liberating hydrogen gas and evolving heat, thereby producing a serious fire and explosion hazard. Some chemicals in contact with other materials will generate heat and give off flammable gases or react explosively. Others, upon decomposition, may generate heat and ignite spontaneously or support combustion through oxidation. A key fact to remember is that although a chemical may not be flammable by itself it may cause fire under certain circumstances. To prevent this, chemicals should be stored compatibly, that is, with other chemicals or materials with which they can live without reacting violently should a leak or other damage occur which will bring them into contact. This same separation of incompatible materials applies equally to acids, oxidizing agents, and the total range of hazardous substances.

Acids. Certain strong oxidizing acids should be isolated to prevent their mixing with other stores. Such mixing may create an extremely hazardous condition involving fire or explosion. For example, perchloric acid should be isolated from acetic, citric, nitric, and sulfuric acids. Special properties of individual acids should be taken into account when designing storage locations and establishing storage procedures. Examples of such properties are the ability to cause nitration, the ability to form dangerous mixtures with other chemicals, and the reaction to water. Nitric acid forms explosive compounds with most organic materials. It forms flammable compounds with nearly all oxidizable material, and some of these compounds are subject to spontaneous ignition. Fumes involved in a nitric acid fire are exceedingly toxic. Concentrated sulfuric acid chars wood, cotton, and vegetable fibers, usually without causing fires. Fuming sulfuric acid usually causes fires when in contact with these materials. The addition of water develops heat which may be sufficient to cause a fire or explosion.

Oxidizing Agents. These chemicals will decompose readily under certain conditions to yield oxygen, and such conditions may be elevated temperatures or contact with other chemicals with which they readily react. Examples of inorganic oxidizing agents are the chlorates, perchlorates, peroxides, and nitrates of barium, sodium, potassium, strontium, ammonium, etc. Organic oxidizing agents such as nitrobenzene are often violent explosives and special precautions must be followed for their storage and handling. Certain oxidizing agents in the pure state present a fire hazard, but because of their ability to furnish oxygen the hazard is magnified and violent explosions may occur when they are mixed or contaminated with even small quantities of certain carbonaceous and combustible materials such as wood, paper, metal powders, sulfur, etc. The violence of reaction depends upon the degree of contamination or confinement and the type of initiation afforded.

One of the best ways of preventing undesired re-actions from chemicals or other hazardous materials is to have a trained specialist, in-house or other, survey the materials used in specific operations, as well as how they are used. He can suggest preventive control measures and also provide informational material to management and employees for use and control of those materials.

While adequate ventilation is a control method for preventing the buildup of heavy concentrations of flammable/explosive vapors in industrial operations, it will also help prevent the accumulation of air contaminants which may cause personal injury, industrial disease, or lead to death. Industrial work by itself may generate hazardous dusts, fumes, and dangerous gases from the materials used in operational processes. Because of the hazard of these industrial byproducts to personnel, property, and the environment in unsafe concentrations, it is essential that they be safely removed from the working area and controlled. Ventilation may be used to rid work spaces of undesirable contaminants when other methods of protection have failed. The particular controls for any industrial operation will depend on the physical and chemical properties as well as the quantities of hazardous materials present, methods of operation, construction of the plant, and other related factors peculiar to specific industries. To insure against a dangerous industrial health environment it would be logical to have a survey performed by a professional industrial hygienist.

One of the materials currently under the spotlight of OSHA is vinyl chloride, which is used in some areas of the plastics industry, and also as a gaseous propellant in some spray containers. It is natural that OSHA has established this material as a target area since it is not only toxic and a fire and explosion hazard, but is under investigation as a carcinogen.

Recent studies prepared by and for the government have indicated that concentrations of the material may be cancer-inducing. Intensive research is currently being performed to determine the degree of concentration that is hazardous.

The foregoing explanations and precautionary measures offer a limited insight into the range of the problems involved in the use of hazardous materials in industry today. In what follows we shall consider some specific situations, and also try to show how hazardous materials may have an adverse effect on other related industries.

SPECIFIC SITUATIONS IN REPRESENTATIVE PLANTS AND INDUSTRIES

The following is a partial listing of some industries or operations with which I have been personally involved. One common denominator they have is that almost all of their operations involve flammables, gases, chemicals, acids, paints, fuels, paint-spray operations, degreasing and other cleaning units, and the entire gamut of solvents, in one way or another. Though the problem areas identified are only stated in general terms, the complexities of related controls are often manifold. For that reason, it would be foolhardy to attempt to provide simple, trite answers. There are none. It cannot be overstated that each circumstance should be explored by itself, by trained specialists, to insure full protection and compliance with existing standards.

Aerospace Industry. Because of gasoline, jet fuels, or other volatile flammables, fire and explosion are an ever present hazard aboard aircraft. In order to guard against the danger of ignition caused by static electricity, aircraft must be effectively grounded during refueling, servicing, maintenance, and other airfield activities. The severe restrictions against smoking near or on board parked aircraft is common knowledge to air travelers and visitors to airports. These same fire precautions and respect for aircraft fuels also apply to flammable materials in manufacturing plants or elsewhere.

Battery Manufacturing and Maintenance. Acids used in the electrolyte, particularly under freezing conditions, create caustic burn hazards to personnel. Moreover, the hydrogen gas generated as a result of electrolytic action may explode when exposed to a source of ignition. Adequate ventilation is only one of the complex controls required.

Blast Cleaning, Buffing, and Sanding Operations. Abrasive and toxic dusts are not only capable of causing external physical injury, but may be absorbed or inhaled into the body. Ventilation, exhaust devices and controls, and approved personal protective equipment are among the intensive controls needed.

Brass and Copper Industries. Moisture contained in porous zinc can induce "pot explosions" (molten splash and geysers of molten metal) when the zinc is introduced with other metals into the molten mass within casting pots or furnaces. One of the many controls is to store the zinc ingots in comparatively warm and dry surroundings for sufficient time to permit drying, prior to being fed into the furnaces.

Cable Industry. Toluene, previously discussed, is often used as a solvent and binding agent for plastic coatings. One frequently sees toluene here and there in cable establishments, together with quantities of MEK which is sometimes used as a cleaning agent for machine parts. Again, as previously pointed out, these materials are extremely volatile.

Electronic and Electro-mechanical Assembly Industries. Flammable solvents of various types are frequently used for shop cleaning of delicate electronic parts and assemblies. Ignition of vapors from these materials may be caused by smoking as well as electrical contacts generated when the units are activated. In addition, one often observes test units in operation without any warning signs or indications that they are activated. The use of simple red light warning fixtures (or bulbs) connected to the unit controls or power source may provide adequate warning to employees and prevent electric shock.

Explosives, Ammunition, and Armament Industries. The very nature of manufacture, storage, assembly, and other use of these products, as well as the testing (and weapons firing) required in these industries demands extensive, specialized, professional controls. Reactions are almost always violent, whether in manufacture or other situations where sparks or static electricity may create the ignition factor. The matter of construction and controls for firing ranges is also an area for special consideration by technically competent specialists.

Fiber Glass. Small particles of fiber glass suspended in the atmosphere or otherwise handled during cutting of this material or other fabrication processes may be inhaled (with consequent pulmonary damage). The results may be quite severe. Minute particles imbedded in the skin may cause infection. Process control equipment and approved personal protection are the beginnings of the total protection required.

Garment Industry. Naphtha and other flammables are still quite frequently (and regrettably) used as dry cleaning solvents in shop areas. The open containers, no matter how small, may create a local vapor buildup in the flammable range, and, in conjunction with smoking or any other ignition source, may result in violent fire. Spillage of this material may be absorbed by fabrics, which in turn become intense fire sources (or wicks) within themselves. Nylon and certain other synthetic materials, as well as woolen fabrics, may create static electricity which in turn can initiate the fire, even if smoking is totally eliminated from the establishment.

Metals and Heat-treating Industries. The storage and in-process use of flammable gases, solvents, caustics, and acids, individually or in conjunction with the heat generated by process operations, has the capability to cause fire/explosion/toxic hazards.

Optics (lasers). The concentrated light beams from lasers, or reflections from glass or other shiny surfaces, may induce severe eye damage. Direct exposure to the end beam of lasers may produce burns of varying severity or even prove fatal.

Paint Industry. The masses of volatile raw materials and finished products used in the oil-based paint industry, as well as in in-process operations, mixing, general handling, containerization, and storage, pose intense fire hazards which in turn call for intense control measures.

Plastics Industries. Vinyl chloride, previously discussed, is one of OSHA's target materials. Since there are constant new developments concerning this material, it is recommended that contact be maintained with OSHA offices (appendix A) to secure rapidly changing and updated information.

Printing and Print Reproduction Industries. Some of the solvents used for inking or reproduction processes, or for the cleaning of presses and reproduction equipment, may have toxic or flammable potential.

Research, Development, and Chemical Laboratories. The complexity of hazardous materials concentrated in one or several areas makes necessary the individual containment and protection of these materials. No doubt the chemists and researchers who use these materials are more cognizant than other people concerning the dangerous effects of miscellaneous chemical and other reactions in those laboratories; nevertheless this same bank of technical knowledge must be backed up by proper housekeeping to prevent creation of toxic and fire problem sources.

Rubber Industries. Many problems arise concerning the use of toluene, acetone, flammable adhesives, and other toxic substances, which often give off flammable adhesive vapors. A mere spark may wipe out an entire plant. Also, the vapors alone may be dangerous to health. The most intense health protection and fire prevention controls should be employed throughout these plants.

Shipping and Longshoring. Almost anything that is shipped in this country is likely to pass through one of our ports. Though flammable and hazardous materials are generally carried as deck cargo, one can appreciate the difficulties which may be encountered either on deck or in an enclosed hold of a vessel upon breakage or rupture of containers carrying these materials. Moreover, all of these materials must be stored on piers or in warehouses before loading or after unloading. This requires storage know-how, chocking and shoring, and much expertise in materials handling both ashore and afloat.

Spray Painting. The matter of spray booths has been discussed. Whether this activity is the sole product line of a company or whether it is one of the process operations used in a major segment of industry today, the very nature of the concentration of raw materials and vapors demands attention. Sources of ignition should not be permitted in areas near these operations. Electrical equipment, exhaust fans and the like, and ventilation must be in accordance with required standards.

Welding. Most welding processes use acetylene and oxygen. These materials are ignited at the controlled nozzle by a spark. If they are accidently released in quantity into the shop atmosphere there is danger of fire or explosion. All compressed gas cylinders should be properly secured when they are in working and storage areas, and properly stored in approved locations in accordance with existing standards. Should a cylinder be dropped and its valve sheared off, the cylinder then becomes unforgiving.

POLICY GUIDELINES FOR MANAGEMENT

A summary of some selected guidelines that management might incorporate in policy statements covering control of hazardous materials follows:

1. Substitute inert materials for hazardous ones wherever possible.

2. Conform to existing standards covering the materials and processes involved (see appendixes).

3. Have in-depth surveys performed by specialists.

4. Prohibit smoking wherever flammable materials are stored or used, and enforce this prohibition.

5. Isolate mass storage and maintain compatibility of materials, preferably at a safe distance from the plant, if at all possible.

6. Insure good housekeeping.

7. Insure positive identification of all materials (flammables, toxics, chemicals, acids, gases, etc.) as well as the containers or tanks in which such materials are used or stored.

8. Insure that minimum quantities of hazardous materials, within operational limitations, are used in process operations for the minimum time required. These materials should not be stored in shop areas.

9. Insure that appropriate approved personal protection equipment is procured and used.

10. Insure that adequate, properly designed ventilation is properly maintained.

11. Prohibit filling, emptying, or transferring of hazardous materials from one container to another within the plant.

12. Insure that no flame, spark, or static-producing materials, tools, or equipment are used in proximity to flammable materials.

13. Coordinate with local fire authorities and solicit their fire protection survey and advice. Provide them with a review of storage and in-plant use of flammables and other hazardous materials. These people will respond to any fire or emer-

gency and it is important that they be alerted to any potential problems in the fire-fighting process, for their own personal safety.

These broad control areas and selected policy guidelines may provide management with some baselines for establishing firm positions to insure safe operations as far as hazardous materials are concerned. To be effective, the policy statements should also indicate:

1. *The company's goals* in achieving the best possible protection.
2. *The methods it will use* to achieve maximum safety consistent with efficient operations.
3. *The personnel responsible* for carrying out these policies.

Once the policy has been written with logical, enforceable safety rules, it should be given the broadest distribution throughout the organization. At a specified later date a report should be submitted to top management indicating the actions which were taken to turn policy into fact.

PLANT DESIGN FOR SAFETY

Most safety problems are caused by people. Many modern facilities, well-designed and constructed, often deteriorate due to an assortment of the practices mentioned above, simply because methods of control have not kept up with the technology of architectural design and construction.

If you are in the fortunate position of designing a plant to meet your own specifications, or moving into an already existing modern facility, consider having your managers and architects "design out" the problems mentioned above. Demand that the requirements of OSHA, NIOSH, the Environmental Protection Agency (EPA), as well as local mandatory regulations, plus current advisory standards, be used as guidelines in the planning process. Consult your insurance carrier, trade associations, ANSI, and the fire department, and draw on any specialized professional consultive assistance which may not exist in-house. When this has been accomplished, your own policies and direction will set the pattern to maintain the conditions that you desire.

Several years ago, the author visited a company in its partially completed new structure. It had recently moved in and was in the process of setting up. It was obvious that the company had growing pains, even before beginning operations. The president asked if I would make an intensive personal survey for him to determine any potential problems that may have been overlooked so that he could have them corrected before construction was completed. Though the structure was reasonably well designed, there were several obvious and major flaws in the basic planning. It appeared that the architects and engineers hadn't fully coordinated with all segments of management, or been provided with well-rounded professional assistance to eliminate some major hazard areas. Full advantage had not been taken of the ample real estate available to the plant, with the result that everything was centered under one roof.

I submitted several major recommendations. The one which was of greatest concern to me was the planned area on the main floor of the plant which was to be used for storage of great quantities of paint, acids, volatile flammables, cyanides, corrosives, and the like. No arrangement had been made for separate storage of incompatible materials. The plans called for storage of all items in one large open room. The design of this room included heavy reinforced concrete walls, fire doors, fire extinguishing systems, venting controls, deluges, drains, and the like, all of which are needed to provide basic protection. Since so many controls were being built into this room, it was also planned to place the stripping and plating operation in the same area. The company had given itself not only a potential storage nightmare, but was introducing the human element as well. Add to this the fact that the room abutted a loading platform which was to contain bins for rubbish and trash for ultimate collection by a contract company, and you have the potential of a truly ultimate nightmare. Internal or external ignition could severely injure the employees in the plating area, could possibly totally dislocate operations, and could possibly put the plant out of business altogether.

Separate inquiries revealed that the actual quantity of flammable, volatile, etc., materials used daily was negligible. The company purchased these materials in bulk quantity to reduce costs. The solutions were obvious:

1. Relocate these materials outside the main plant at a suitable safety distance, in a lesser structure of their own, protected within existing standards.
2. Arrange for compatible and separate protected storage of such needed materials.

3. Withdraw only the quantity needed for daily consumption.
4. Set up a separate protected plating/stripping room.
5. Relocate the trash collection area away from the main building.

All of which would result in:

1. Less exposure to plant and personnel.
2. Reduction of hazard to main plant.
3. Drastically reduced insurance rates.
4. Release of interior plant floor space to manufacturing operations.
5. Peace of mind to management.

At the conclusion of my survey, I requested a meeting with the president, only to be advised by his secretary that a board meeting chaired by him was to start momentarily and couldn't be held up. Since I had a long way to travel, with my own tight schedule, I requested that she slip him a note, in which I wrote, "Desire 5 minutes, may offset major problems and save much $." He saw me immediately. I presented my findings and recommendations, he concurred wholeheartedly and told me that he would save more on insurance rates alone than I had immediately calculated. The president requested that this information be presented to his staff, inasmuch as they were already in conference. During the presentation, I pointed out that besides the reduction in exposure and hazard to plant and personnel, and the financial savings involved, the relocation would make unnecessary the massive internal protection that had been planned for the storage segment and would release valuable internal real estate for adjacent operations which already appeared to be crowded. The manufacturing manager breathed a sigh of relief and said that he was hungry for space to be able to expand a new product line.

The proposals and recommendations were adopted. Modifications and changes were subsequently made, and the lowered insurance rates and operational benefits not only covered the costs but put money in the bank for a long time to come. The chief executive redefined the future safety policy of the company and set the pattern for a total systems approach to all future planning and operations.

Here was a case where management did not originally intend to cut costs by eliminating necessary protective measures. The total planning process had not been sufficiently brainstormed by all responsible managers; the company had neglected to take advantage of outside specialists; and too many safety standards had been applied as an afterthought. Closer coordination with the insurance carrier, the fire department, the trade association, etc., would have helped to prevent most of the company's plant design problems.

Industry generally has the capability to plan its operations for both economy and production, and to a large extent it has the know-how to manage such operations with efficiency. A company's success in avoiding catastrophic loss depends on top management, however, and the way in which management's policy is carried out by all elements. If all levels of management and supervision are made fully aware of it, not only for plant internal operations but also in the total planning process for plant and structure (new and old), accidents and the breeding grounds for accidents will be reduced.

A plant is a complex of people, material, and machines. Since industry must care for the machines to prevent breakdown and down-time, by the same token industry must evaluate all efforts to care for the people who run those machines. If it does, the work force will stay, and a stable group of trained personnel will accumulate its know-how. Employees will travel a few extra miles because your plant is a good place to work.

8

The Safety Manager

Call him what you will — safety director, safety
engineer, safety officer, loss prevention specialist — he
is your representative delegated to direct your accident
prevention program. It is not his responsibility to run it.
That is the responsibility of the top manager, because
running an accident prevention program means direct-
ing the activities of all other managers. The *implementation* of a safety program rests with all levels of
management and supervision.

Too often, some nonprofessional corporate representative is given one of the titles above, and thereby
assumes the activity of the safety specialist in addition
to his other duties. A questionable policy, *unless* the
organization is so small that there are a mass of over-
lapping functions and you simply cannot avoid doing it.

Too often, also, in a very large company this same
individual is burdened with other activities which
fragment his efforts and diminish the time and planning
available for his safety assignment. Too often he is
bogged down with basic personnel actions and com-
60

pensation reports, necessary functions, but functions which can be performed by other personnel. An accident has happened and the situation which existed cannot be restored. The safety manager should investigate and use his technical ability to assist in determining the basic causes of the accident, but it is questionable whether his time and technique is used to best advantage to accomplish the consequent clerical tasks.

As pointed out earlier, a professional safety manager may reasonably be expected to pay his salary back to the company a considerable number of times over in savings generated through his efforts. His primary effort should therefore be directed to preventing accidents, to planning a suitable safety program to fit the enterprise, and to coordinating the activities which translate this program into action by the responsible managers.

A safety director must have the direct backing and support of the company's top executive. Since a safety director crosses all organizational lines, coordinates with all levels of management and labor, and treads all of the complex paths from the executive and staff offices to the workplace, he can report to no specific level of management other than the top manager. The latter person is the one who bears the final responsibility and accountability, and who must answer to corporate headquarters or the stockholders (depending on his position), and himself, if he incurs any deficits through avoidable accident losses.

I have never been able to understand why the safety director is so often placed on the staff of the personnel director, the director of industrial relations, or other

staff managers. In my opinion, to place a safety director in this illogical position, or somewhere else down
the corporate ladder, not only undermines his authority
and creates another questionable level of management
between him and the top executive, but also tends to
mirror the company's attitude toward safety. To
prevent a safety director from sitting in on meetings
of top management and staff is to prevent him from
directly providing and coordinating decision-making
information with these people. The result is a diluted
safety program. The U.S. Army, and many of the
largest and most renowned U.S. corporations, long ago
established the safety director as responsible to the
commander or executive manager.

Independent surveys concerning the organizational
placement of the safety professional reveal some
significant and important information regarding the
impact of such positioning in the corporate management structure. The National Safety Council reported
on a survey of safety professionals in American and
Canadian industry:[1]

> Of the respondents, 44.8 percent reported directly to a
> member of top management, such as the president, vice-
> president, general manager, finance officer, or works man-
> ager. Another 19.8 percent reported to the plant manager or
> superintendent, and 30.4 percent reported to the industrial
> relations director or personnel manager. The remaining
> 5.0 percent reported to the managers of various depart-
> ments (labor, insurance, security, medical, etc.) or to the
> chief engineer. Importantly, another survey, conducted by
> the American Society of Safety Engineers, showed fre-
> quency rates and severity rates of disabling injuries are

---

[1] National Safety Council, *Accident Prevention Manual for Industrial
Operations*, 7th ed., 1974, p. 58.

considerably lower in those plants where the safety professional reported directly to a member of top management.

The National Safety Council survey was conducted in 1961, and the results may or may not reflect the current safety professional's position in industry. There is no evidence that the picture has substantially changed since 1961.

My comments concerning the placement of the safety director in an organization's structure are in no way intended to reflect on the importance of the intermediate directors or managers to whom many safety managers currently report. The comments are an attempt to put the safety position into its proper focus. In my opinion the position should be an independent, specialized staff placement, with heavy corporate impact, much the same as legal counsel or comptroller, and it seems to me to require latitude of action and authority to function to its best advantage. I have had business relationships with more than 200 companies, ranging from corporate giants to the smaller independents. These direct associations with responsible executive managers usually resulted in mutual open candor. The substance of a great portion of this book is the distillation of that exchange process. I would like to cite three specific examples from personal experience of problems which arose from structuring the safety director under intervening layers of management. In each of these instances, all involving major divisions of "Fortune 500" companies, the managers were mature executives for whom I had the utmost regard. In each instance, that executive would generally take decisive action in resolving major problems that arose, once I had presented the facts of the problems to him, together with concrete recommendations for resolution of those problems. We'll call the corporations *A*, *B*, and *C*.

Corporation A had a young, inspired, but green safety director who reported to the director of industrial relations. In addition to his varied other duties, he truly desired to learn the scope of his safety assignment, and to explore all of its facets within his capabilities. Though he was a novice, he caught on rapidly and began to amass what knowledge he could in that area. He realized his limitations, but when he knew that he was right he would not hesitate to take a firm position and clearly present his views. His boss was a proficient prime mover and normally would not hesitate to take corrective action and resolve matters in conjunction with the top-tier executives concerned. He was demanding, but also action-oriented. Here was an excellent combination: a strong, capable top executive, and a sharp subordinate.

There were problems, however. There always are. Often in his rounds the safety director would make recommendations and requests to middle managers or line supervisors. In too many cases, inertia reared its ugly head and little understanding or cooperation was shown and no corrective action was taken, despite the fact that some potentially catastrophic hazards existed. The formal and informal organizational structure and lines of communication then came into play. Safety would report to his boss, and the opposing manager would often use every effort to block the matter and get his personal views and purposes upstairs to *his* top manager. The strong director of industrial relations (when firmly convinced that his subordinate was correct) would take the case to the responsible executive, and in many cases the problems were resolved. In several cases they were not, however, for one reason or another, and only rarely did the matter get to the top for resolution. Severe problems remained problems, *and the chief executive knew nothing about them.* Nevertheless, the safety manager hung on and learned, and because of his immediate boss's general support and action, he began to get some things done. He appreciated his boss's support and responded accordingly.

In this instance, lack of direction resulted in the safety office fighting brushfires rather than managing a broad and planned program. In some instances the safety director may have been too abrupt in presenting his initial recommendations to middle managers, or line supervision, and thus he met with psychological resistance. In some instances there may have been a conflict in ideologies between the director of industrial relations and other executives involved. Some serious cases should have been taken topside for final resolution only when they could not have been resolved at other levels of management. But all are instances which result in massive frustrations and delay, as well as in hazard continuation and exposure. Here is a case of personality clashes overriding important problems — with top management kept in the dark!

The director of industrial relations was subsequently promoted to a vice-presidency. His successor lacked aggressiveness, and hesitated to "rock the boat." As a matter of fact, it appeared that he didn't want to be informed of problems by his subordinates, or to face the fact that there were any major safety (or other) problems in the organization. He certainly hesitated to take conflicts upstairs. Without support, the safety effort began to falter. Result: The safety manager resigned and got a better job. A weak successor was appointed from within the personnel division, a person without any previous significant training in safety. The corporation suffered.

The very nature of the company's naturally hazardous product line further warranted a strong product safety group. Whether or not it existed as an integral unit, I never found out. If it did, it must have been well hidden. In my many dealings with this company, I felt that efforts in this particular area were disjointed, and that safety matters were treated only on a crisis basis.

Company A is a massive independent division within its corporate enterprise, with huge domestic and foreign sales. It warranted a strong, independent, professional safety manager to provide it with a bank of safety knowledge, continuity of operation, and corporate placement which would permit that impact value to be used to best advantage. I believe that the savings in insurance rates alone, generated by this manager in coordinating action to eliminate the glaring weaknesses in the company's safety posture, would have been more than enough to have supported a professional safety *staff*.

Company B had a different organizational structure. A significant part of the division's organization chart looked as shown in Figure 1. Corporate safety policy was strong. The divisional executive manager was

FIGURE 1
Company B's Organization

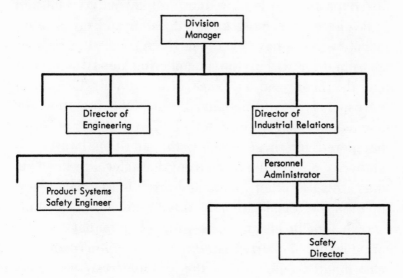

dynamic and would not hesitate to call an immediate meeting with his responsible directors, and seek an immediate solution, when a serious problem required resolution. The director of industrial relations was also topnotch. His personnel administrator was a likeable, efficient, and energetic subordinate, whose occasional failing was that he sometimes hesitated to bring sticky matters to his boss's attention, even though they couldn't be resolved at his level. The safety director, however, was locked down below. Further, he was a "combination man" who was saddled with other responsibilities, despite the fact that this major division was a multimillion-dollar separate enterprise with several satellite manufacturing/operational plants and activities reporting to it—none of whom had a separate safety manager.

Product/systems safety was vested in the engineering department. The product/systems safety manager was one of the most knowledgeable and cooperative people whom I have ever met. Regrettably, higher-level coordination within the engineering department was not the best, and at times it was totally lacking or unresponsive. Additionally, besides the fact that the competent safety director's efforts were severely hampered by these factors, the additional layers of authority above him hamstrung his efforts in his dealings with the other plants and directorates. His placement in the organization too often reflected the attitude shown him in his dealings with other managers. The organizational chart created roadblock upon roadblock. Coordinated effort with the product/systems safety manager was extremely limited, and was again hampered by the intervening levels of management. Since the very term "systems safety" (which will be discussed in Chapter 11) includes all elements in the life-cycle of a product from concept through design/ engineering, manufacturing (including quality control, packaging, and shipping), and marketing, we can appreciate the levels of authority which intervene. We can further realize that product/systems and industrial safety must work hand-in-glove with all other managers to insure that the integrity of the product (and its safety characteristics) is insured throughout that life cycle, most heavily on the production line.

A solution to this problem would be to have an independent safety office reporting to the divisional manager, manned by a strong safety director, and a product/systems safety manager. In the case of Company B, it is my opinion that the scope and nature of its sophisticated and hazardous product line and activities, as well as its total sales impact for the corporation, would have warranted such a structure.

Company C had a different set-up. The organizational structure and position of the safety staff was as shown in Figure 2. The director of industrial relations was another capable executive. The manager of the medical department (a doctor) was an intense, kindly, and very astute man who was engrossed in running his medical department, as such. The safety office was manned by very able people, but it appeared to be a disjointed operation. I could never figure out why it was structured within the medical department, particularly since the very technical nature of the massive company warranted intensive cross-coordination between the safety people and its operating elements. Whatever the reasons, it will take a lot of convincing to ever get me to support positioning the safety office under the medical staff in any organization—with no reflection

FIGURE 2
Company C's Organization.

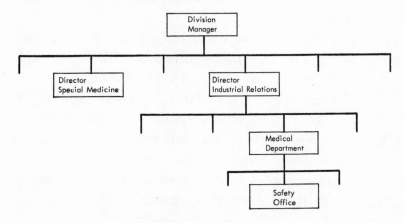

at all on medical staffs. It just seems to me that the two organizations should be separately run.

The nature of company C's product line was sophisticated, covered a broad expanse, and was heavy in research areas which further warranted a director of special medicine, who happened to be one of the most renowned and best in his field. I was never aware of a product safety group, per se, within the division. Although all of the individuals mentioned above were of the highest caliber individually, I felt that the total approach to safety was loosely tied together. I could understand the safety department's need for coordination with the medical department, but could not see it reporting to that manager. The chief of the medical department hesitated to bring major unresolved safety problems to the attention of the director of industrial relations. The latter was beset by massive problems in other areas, and as a result did not display any major interest in the safety activity until a crisis loomed. He relied almost completely on the medical department. His inaction fostered an inability of the safety people to get the results that they might have gotten otherwise. Other clerical activities consumed a good part of their time with minutiae.

The combination of these circuitous layers of management, as well as the requirement of the safety staff to audit the safety activities of several dispersed manufacturing/operating plants, without authority to enforce its requirements, resulted in frustration of that staff. Fortunately, however, the various managers and executives of the division were capable and reduced most problems within their areas to a minimum. Still, there were severe problems. The division and its corporate parent are two of the most respected enterprises on the national and international scene. Corporate safety policy is strong and the corporate safety office—amazingly—is structured differently. Be that as it may, I believe that a divisional safety staff, with authority and latitude of coordinated action and reporting to the division manager, would be much more effective for that organization. Massive corporate (and division) sales would sustain it.

To be effective and responsive, a safety manager should make a thorough study of the structure and operations of the corporation for which he works so that his program is reasonable within the demands of the organization. At the outset, the new, or inexperienced safety manager should:

1. Learn the company's blueprint for accomplishment of its objectives.
2. Review management actions, as reflected in directives, reports, and so on, and its attitudes toward support of an active safety program.
3. Review operational methods and basic planning.
4. Become familiar with the plant and working environment.
5. Review employee/supervisory attitudes toward safety as reflected in work habits and procedures.
6. Research previous first-aid and disabling injury reports to uncover trends and problem areas, determine causes (and causes of causes), and review corrective actions. An analytical review of every accident report will indicate the degree of investigation which was conducted, and reflect the attitude of the supervisor/manager who completed

and reviewed it. One of the most dismal results of an "investigation" is something like this:

"Cause of accident: Employee was careless. Corrective action: Told employee to be more careful."

7. Research property, equipment, and material losses attributable to accidents.

Once having researched these major areas, the safety director is then in a position to analyze and evaluate where further major efforts must be directed. Priorities must be determined based upon the seriousness of the deficiency. Having done his homework, he must plan a basic program, staff it with responsible management levels for coordination, and present his course of action and program to his boss. The program must be based upon sound, realistic reasoning to meet the requirements of the company, and should include cost figures for accomplishment, within reason (that is, the ratio between incurred losses and cost of projected requirements).

Once having defined and documented the program, and had it approved, he must then present it to all staff managers. They are the ones who will put it into action. The safety manager should stand ready to assist, advise, and help in any way, but he must never attempt to run the show by himself. If he attempts to do so, a natural reaction will set in at all levels of management and supervision. They will fall back and let him do everything himself. He will accomplish nothing.

Should the safety manager move into an area, take over, and issue orders—thereby completely bypassing the supervisor in charge of that area—he will generate resentment by that supervisor and others. He can then expect little cooperation. No one wants to be bypassed, and management levels must be respected if the safety manager desires to succeed.

Too often, safety assignees run off in all directions with a program which is going to accomplish wonders. A big splash is made about trivia, and wheels are spun in the areas of least concern. The main problems receive casual attention. Everything cannot be done yesterday, of course, and determined effort and study must be made to consider the facts, determine the course of action, and then execute the action. But possibly the worst thing that a safety person can do is to hang his hat on a cliché and play trivia for all it is worth. There is no single solution.

The safety manager does not stop accidents. He provides information and guidance to top management and supervision so that they, being in control of personnel and facilities, can take appropriate measures to prevent the build-up of circumstances which can result in accidents. People are fallible. Therefore, it is necessary to put into practice an action program which will optimize reliability potential.

Since managers are responsible for pursuading people to work to required standards, the focus of effective safety performance must be on management action. Management is always interested in safety, but we are here concerned with the *degree* of interest, specifically that which is manifested in action.

The need for management action therefore focuses on the highest echelon, the top man, rather than first-line supervision. Safety managers should develop decision-making information for principal managers, rather than personally assuming the responsibility of promoting employee awareness.

Since responsibility for total safety performance rests with the executive, the safety manager should logically be his staff representative, and must be prepared to provide facts to assist the top manager in triggering direct action. Responsible management desires facts which are generated through dynamic action.

Here is a note of importance which you may want to provide to the newly appointed nonprofessional safety manager:

"Secure every bit of training and keep learning.

Ferret out and join the local chapter of the American Society of Safety Engineers, and other local safety organizations. Many friends will only be too glad to pool their knowledge, and provide all of the help that they can.

Secure the curricula of the National Safety Council, universities, and colleges which provide rounded courses on safety management.

Accumulate a library.

Start off with formal courses in safety management, and go on to courses in industrial safety.

When this background has been secured, proceed to specialized courses for our own industrial operations.

Trade associations will help.

Our own insurance carrier will provide necessary assistance, and will no doubt assist in launching our program.

We are in a dynamic field of safety endeavor, one which requires personal dynamism to keep abreast of it."

Do not expect to stop all, or the majority of accidents, and accomplish everything overnight. The translation of safety control theory from a routine industrial environment to a near-optimum safety program takes time, patience, and tenacity.

9

Basis of Accident Prevention

A fundamental analysis of why accidents happen demands a logical approach. Essentially, an accident is an unplanned event which has the capacity to cause injury or damage, and is attributable to either an unsafe act or an unsafe condition. Several generally accepted theories set forth by H. W. Heinrich[1] many years ago, based upon sound reasoning and formidable research which bears upon how and why accidents happen, lead to some dramatic conclusions concerning the accident sequence and ratios.

THE ACCIDENT SEQUENCE

The Domino Concept

The psychology of how and why people do what they do involves their background, their social environment, and the circumstances which molded them. Also im-

---

[1] Herbert W. Heinrich, *Industrial Accident Prevention*, 4th ed. (New York: McGraw-Hill Publishing Co., 1959).

portant is the attitude of the individual, or his acceptance of the right and wrong way to do anything, whether
it be on the job or at home. Too often, the word "fault"
is used as a way of explaining errors that persons make,
without considering either of the foregoing factors. In
addition there is the unsafe act or unsafe condition
that immediately precedes the accident.

If these dominos are lined up in that order, and any
one of them is knocked down, then others will fall also.
Remove any of the dominos, or factors, between the
first and the last, and a gap is created which precludes
the final occurrence. The key to accident prevention
is to remove an intermediate domino.

It follows that one of the vital factors is to *break the
chain of events*. There is little that we can individually
do to change the *social environment* of an employee before he reaches our employ, or even after. This is usually beyond the scope of the single employer or company. Sometimes one or several companies or associations may exert a heavy impact on a community, and
their community relations efforts, coupled with local
and state government efforts, may have a long-range
effect on the well-being and common outlook of the
community.

Attitude must be generated within the company, by
management and supervision, to insure that employees
perform in accordance with acceptable standards and,
more so, that this desire is generated from within,
through effective leadership. The will to accomplish
may only be fostered by reasonable application of every
principle of management, by the employee's acceptance that he is a true member of a team—and not
merely an unknown cog—in the overall enterprise.

Unsafe acts or unsafe conditions can be controlled by eliminating them through research, training, application, and supervision. Once the unsafe act or condition is eliminated, or reduced to its minimal chance of occurrence, the accident is then precluded.

The layman may consider some accidents to be "minor" in their potential for injury or damage, and others to be more severe. Professionally, however, *all accidents must be considered to have serious potential.* It's often a matter of luck, or a fraction of an inch or a second, which makes the difference. The total accident spectrum encompasses not only injury to personnel, but also damage or destruction of property or product, or production of needless scrap or waste.

THE RATIOS

Some years ago, Herbert W. Heinrich concluded that if an improper act was performed 330 times, the end result would be that in 300 cases nothing would happen. In 29 cases there would be a minor, or nondisabling accident. But one of the cases would result in a disabling or major accident. What isn't known is when the disabling injury will occur—the first or the 330th time?

Over the years, Heinrich and other professional safety people used various other figures for the total amount of times that the improper act would occur before resulting in the major accident (or even the "minor" ones), some even concluding that an improper act might occur 1,000 or more times before a misfortune. (And, of course in some extremely hazardous industries, the unsafe act may only occur a very few times before the serious reaction.) The ratios mentioned, are only averages.

Whatever the case, it follows that if any task, or activity, is improperly performed, it is only a matter of time (or luck) before the major accident occurs. As for the "minor" accident—*consider every accident to have potentially serious consequences.*

It was also concluded that in an average of all accidents which were analyzed, 88 percent were attributable to unsafe acts of people, 10 percent to unsafe conditions, and 2 percent to Acts of God. Though all professional safety people do not always accept such ratios, the fact remains that the major cause of accidents is the way people *do* things. The lesser percentage of accidents resulting from unsafe conditions warrants further discussion later. The category of Acts of God also warrants discussion. Any master of a vessel must know how to handle his vessel under varying conditions, whether they be typhoon, heavy swells, hurricane, or something else. A farmer in the Midwest prepares a storm cellar to be secure from tornado. His know-how therefore offsets the accident potential of Acts of God.

Though most accidents are due to unsafe acts of people, the common sense of fundamental accident prevention demands that unsafe *conditions* be corrected as an initial priority. Correction of these conditions then reduces unsafe exposure and reduces the number of unsafe acts.

Whatever the case, we too often chase the physical condition, the "thing" that causes the accident, rather than concentrating on the prime source of the problem —the unsafe acts of people! This is one of the limitations of OSHA.

The immediate costs of accidents are only the tip of the iceberg. The ratio of immediate costs to total costs may be 1 to 4, 1 to 10, or much higher. Close analysis of an accident may even boost the proportion to 1 to 1,000. Incredible, but sometimes true. We'll pursue these hidden costs in depth in the next chapter. The point to remember is that whatever the proportion may be,*the known or visible costs are only a fraction of the total costs.*

CAUSES AND CAUSES OF CAUSES

As we pointed out in discussing the accident sequence, there are several factors or combinations of factors which lead up to an accident. But accident reports usually only list the immediate factor which resulted in the accident—the last thing done, or occurring, prior to the accident. Corrective action is consequently only a correction of that immediate cause. Is this sound?

Let us assume, as a case in point, that a breather tube on an overhead motor pivots due to vibration and continually drips oil on an aisle. For several days the oil patch is not cleaned up, until one day an employee slips in the oil and is injured. The accident report states: "Accident due to unsafe condition, oil on floor. Corrective action: Cleaned up the oil."

Analytically, working back from the immediate cause, we actually have the following sequence:

Oil patch—Unsafe condition

Oil patch not reported—Unsafe acts (of all who did not report)

Oil patch not observed by the injured party—Unsafe act

Oil drip from motor—Unsafe condition

Breather tube pivoted by vibration—Unsafe condition

Breather tube not properly secured in the first place —Unsafe act

Questionable location of motor (or aisle)—Unsafe condition

Questionable lighting in the area—Unsafe condition

Setting up the layout—Unsafe act

What caused the vibration to pivot the tube (mechanical defect or lack of maintenance?)—Unsafe condition(?)

Any action taken to compensate or correct?—Unsafe act

Another case in point: A motorist either fails to get his brakes checked when they are not properly functioning while driving, or he neglects having them repaired when advised to by his mechanic. As a result, he eventually strikes another vehicle or a pedestrian. A superficial accident report might dismiss the matter by stating: "Loss of control, bad brakes, unsafe condition." Had the investigator delved beyond the obvious, the last thing which occurred prior to the accident, and performed a thorough investigation, he would have determined that the *unsafe act* of the motorist in neglecting to have the brakes repaired was the true cause of the accident.

These two cases used as illustrations merely indicate that (1) the obvious is rarely the answer, and (2) thorough investigations should be conducted to determine the root cause(s) of accidents. Executives might then discover any management or other deficiencies which led up to the accident and initiate policy for positive action to prevent recurrence. Since senior management requires thorough analysis of financial and production reports, it should also be provided with accurate and in-depth reports of accidents — which are invariably reflected in corporate finances and production.

To insure accomplishment of such actions and that management is kept fully informed, corporate policy should require thorough and meaningful accident investigations. Demand a complete analysis of reporting of accidents by all involved, so that the basic causes of the accident can be determined and all conditions bearing on the accident corrected. The best technique is the early analysis of factors bearing on the causes of accidents, through a systems approach, prior review of operational procedures, training, etc. Was the operation suitably staffed in the thinking stage, before creation of the situation which caused the accident to come about? If we have not seriously planned our activities to provide for the elimination of accident-causing circumstances, a review is in order. Finally, does the corrective action give a superficial excuse or a well-thought-out solution?

An effective measure employed by many successful major enterprises to trigger action at intermediate management levels is to require that the immediate supervisor complete the accident report within a specified time after the accident. All successive levels of management should then review and sign the report, not only to insure that adequate action had been taken but to comment on any further actions that might be taken to prevent similar occurrences in other activities of the company. This review process will not only reveal unforeseen operational errors or other deficiencies, but will keep intermediate management informed. A final policy requirement is for a copy of the report to be forwarded to the senior manager, besides the normal distribution to the safety director and any other internal corporate administrative areas.

One of the most effective measures that may be used to demonstrate top management's interest is to be placed on the distribution list for a copy of the accident report. The safety director should subsequently provide that same manager with periodic summaries which will permit an across-the-board review of the accident picture. The review may then become an agenda item at the next general staff meeting.

MEASUREMENTS OF EFFECTIVENESS

The generally accepted yardstick of safety-program effectiveness is to consider the accident frequency rate as the measure of the safe working environment of a plant. This rate is based upon disabling injuries, wherein the employee loses at least a day following the accident, and is the ratio of numbers of disabling accidents to man-hours worked. In some cases (although rarely), the company's accident severity rate is computed as the ratio of days lost to the number of man-hours worked.

But is either of the foregoing a truly valid measuring device? I think not. The indicators fail to include any mention of first-aid cases, which were first-aid only because of luck. They further fail to record the noninjury accident, which places the spotlight on property damage, product loss, and related unrecorded deficiencies. No mention is made of damage to plant or equipment, damage to product in course of manufacture, damage to raw material or subcontracted items, prior to entering the manufacturing process, damage to finished product awaiting shipment, loss of parts, assemblies, or subassemblies because of negligent industrial housekeeping or warehousing, rework cost of damaged material, parts, or product, downtime of damaged equipment, cost of cleanup, repair, and renovation, cost of repackaging, crating, shipment, standby time of employees interrupted by a noninjury accident, loss of time by rubberneckers, loss of time by everyone who discusses the accident, adverse customer reaction to all of the above, bearing on delayed delivery, etc. All of these noninjury accident costs raise operational costs, yet most of them are "absorbed" in the accounting for some other activity within the plant. The result is that your true accident costs are not clearly determined, and lopsided operational costs are generated.

Nothing is so heartbreaking to an operations manager than to finally finish a batch of machined parts with close tolerances, while fighting a tight delivery schedule, only to have them ruined by a wrong dip in an unidentified vat in the wrong solution in the plating room. An effective control procedure is to insure that every situation which may be peculiar to your own industry or operation be investigated, documented, and statistically recorded under departmental costs, as well as on the plant's total accident debit report. It's discouraging at first to have all of the skeletons pop out of the closets at once, but such reporting will give you a clear idea of losses which your nonsafety experts may not find, reveal the reasons for these losses, and provide the resource material for preventing similar losses. Not to require such reporting is to accept the fact that the same losses will occur over and over again.

10

Costs

Traditional nonprofessional approaches to assessing a company's accident posture and related costs generally only consider accidents to personnel, as reflected by injuries and fatalities. The obvious and immediate direct costs (medical, compensatory, and administrative) relating to such accidents are included. The indirect costs of an accident, which are generally overlooked to varying degrees, reveal many more dimensions of the organization's accident posture, as well as spotlighting additional expenditures attributable to those accidents — and revealing management or operational deficiencies. As mentioned before, management desires a logical assessment of its operations, and should demand a total audit, indicating all costs attributable to *all* accidents.

Table 1 lists the known costs of an accident, using some of the key items discussed in previous chapters. Table 2 lists the submerged costs. Some categories and costs in the tables are repeated with monotonous regularity because they are situations with which we must

TABLE 1
Accident Costs—Direct

Loss Category	Typical Occurrence	Reflected Costs
Personnel injuries	Fatality Disabling injury Occupational illness First aid	Medical payments Compensation payments Legal/court fees Jury awards

TABLE 2
Accident Costs—Indirect

Loss Category	Typical Occurrence	Reflected Costs
Personnel injuries	Fatality Disabling injury Occupational illness First aid	Administrative expenses Higher insurance liability rate Loss of skilled employee Retraining new worker Lower morale of work force Lost production—new employee Lost time—everyone who discusses accident (including following shifts if the accident was serious) Standby time—employees at same operation, or nearby Investigations—including questioning of all immediately concerned or proximate to the accident Equipment downtime—waiting for insurance representatives, cleanup, teardown and reassemble Production delays Shipment delays Adverse/unsympathetic customer reaction (due to delayed shipments) Adverse dealer/salesmen reactions (due to delayed shipments) Demurrage charges (railroad or ship tie-up)

Property damage

- Damaged forklift or other materials handling equipment (MHE)
- Load dumped by forklift or other MHE
- Improper stacking, warehousing, or operational handling
- Fire/explosion
- Motor vehicle accident
- Damage to production machine or assembly line
- Stacked material struck by forklift or other MHE
- Damage to plant and structure by forklift, other MHE, or personnel accident
- Industrial fatigue induced by improper lighting, noise, cold, heat, dust, fumes, or poorly designed work situation (human factors engineering)
- Exposure to elements
- In-plant exposure to contaminants

- Shutdown of assembly line (depending upon seriousness of accident)
- Repair of equipment
- Replacement of equipment
- Production delays
- Shipment delays
- Cleanup of equipment
- Downtime of equipment
- Loss of parts
- Standby of employees
- Discussion by other employees (depending upon seriousness of accident)
- Rubberneckers
- Investigations
- Administrative expenses
- Reinspection of repaired material (quality control)
- Loss of vehicle for "x" time
- Adverse/unsympathetic customer reactions (due to delayed shipments)
- Adverse dealer/salesman reactions (due to delayed shipments)
- Delay—waiting for insurance representatives if accident was serious
- Demurrage charges (railroad or ship tie-up)
- Disposal of debris

TABLE 2—Cont.

Loss Category	Typical Occurrence	Reflected Costs
Product damage	Dropped from forklift or other MHE	Production delays
	Improper stacking, warehousing, or other operational handling	Shipment delays
		Replacement
	Hit by moving equipment	Reworking or repairing
	Fire/explosion	Reassembling
	Motor vehicle accident	Reinspecting (quality control)
	Industrial fatigue induced by improper lighting, noise, cold, heat, dust, fumes, or poorly designed work situations (human factors engineering)	Repackaging/recrating
		Returning to plant and reshipping
		Adverse/unsympathetic customer reactions (due to delayed shipments)
	Improper housekeeping	Adverse dealer/salesmen reactions (due to delayed shipments)
	Exposure to elements	Demurrage charges (railroad or ship tie-up)
	In-plant exposure to contaminants	Discussion by employees
	Plating or degreasing errors due to lack of (or improper) identification of tanks or vats	Rubbernecking
		Disposal of debris
Scrap loss	Dropped from forklift or other MHE	Replacement of raw material
	Improper stacking, warehousing, or operational handling	Replacement of subcontract items
		Repair of subcontract items
	Hit by moving equipment	Reassembly of subcontract items
	Fire/explosion	Reinspection of subcontract items
	Motor vehicle accident	Loss of needed parts
	Housekeeping	Cost to manufacture lost parts
	Exposure to elements	Production delays

Scrap loss (cont'd)

In-plant exposure to contaminants
Degreasing or plating room errors attributable to improper (or lack of) identification of tanks or vats
Industrial fatigue induced by improper lighting, noise, cold, heat, dust, fumes, or poorly designed work situation (human factors engineering)

Shipment delays
Adverse/unsympathetic customer reactions (due to delayed shipments)
Adverse dealer/salesmen reactions (due to delayed shipments)
Demurrage charges (railroad or ship tie-up)
Discussion
Rubbernecking
Disposal of debris

Undue maintenance (attributable to accidents)

Damage to equipment
Damage to plant by forklifts and other MHE
Slipshod housekeeping
Slipshod machine operations (uncontrolled machine chips and turnings, uncontrolled coolant splashed over floors and equipment)
Improper use of or overload of equipment
Lack of basic preventive maintenance of production equipment and facilities by operating personnel and other employees
Negligence (possibly attributed to employee attitudes)

Cleanup
Repairing/rebuilding
Renovating
Teardown
Disposal of debris
Delay in other maintenance (which will be reflected in additionally induced costs)

cope. An attempt is made to indicate typical occurrences and reflected reactions, but the list is by no means complete. Additional input may be individually made to suit a specific industry or operation.

Since it is obvious that the hidden costs total more than the known costs usually presented to management, management might logically demand a total accounting and reporting of all costs or operating expenses which were incurred in the company's operation. As we said in chapter 2, executives should require a complete breakdown of all costs which were incurred in any disruptive accident. Such a financial accident accounting would include (and we repeat):

1. *Personnel injury costs.*
2. *Production delay costs* due to accidents.
3. *Property damage costs* due to accidents.
4. *Product damage costs* due to accidents.
5. *Shipment delay costs* due to accidents.
6. *Scrap loss costs* due to accidents.
7. *Rework costs* due to accidents.
8. *Maintenance costs* due to accidents.
9. *Miscellaneous costs* due to accidents.

We leave it to the customer relations department to determine the impact on the customer who did not get his product on time, not to speak of the impact on the dealer who is thus embarrassed.

11

Product/Systems Safety

The far-ranging provisions of the Consumer Product Safety Act (CPSA) of 1972 (Public Law 92–573) have not only presented U.S. industry with a new major challenge, but have provided the government with powerful tools for mandatory control of the safety of consumer products, as well as the legal machinery for enforcement of its requirements. The Consumer Product Safety Commission (CPSC) has been established as the regulatory agency for the Act, with broad enforcement powers.

A concise summary of the provisions of the law and the authority of the CPSC states:

> The Consumer Product Safety Commission has been assigned the task of assuring that the numerous products and components covered by the law are safe or safer. It is armed with comprehensive authority to set safety standards, ban products that cannot be made safe, rid the market of imminently hazardous products, require recordkeeping, examine records, call for reports, inspect business premises, impose labeling-warning-instructional information directions, and

demand safety certifications. Add to these the law's provisions covering defect notices, recalls, replacements, repairs, purchase price refunds, information collection and publications, new products, exports and imports. And much of this is backed up by enforcement devices — civil and criminal penalties, injunctions, seizures, condemnations, import exclusions, and consumer suits.[1]

The purpose of the CPSA, as stated in the act, is: "To protect consumers against unreasonable risk of injury from hazardous products."

A consumer product is then defined as any article, or component part thereof, produced or distributed

— for sale to a consumer for use in or around a permanent or temporary household or residence, a school, in recreation, or otherwise, or
— for the personal use, consumption or enjoyment of a consumer in these pursuits.

Many products are excluded from this definition (e.g., motor vehicles, aircraft, boats and certain accessories) since they are separately covered by other statutes. Further, the OSHA and other existing standards cover almost every other type of product produced by industry today.

Industry bears not only the moral but the legal responsibility to insure that its consumer products are tolerably free from hazards in design, construction, operation, handling, storage, shipping, and environmental impact. Combined with these factors, operational and maintenance manuals, labeling and other instructions should clearly point out necessary cautions and warnings which will alert the user to hazards

---

[1] *Consumer Product Safety Act — Law and Explanation*, Commerce Clearing House, Inc., Chicago, 1973.

inherent in use of the product. To do otherwise is to invite customer injury, potential legal actions, and resultant financial liability by the company, as well as adverse reaction of the customers, government, and the public at large.

In order to provide decision makers with an effective solution to insure compliance not only with the CPSA but also the current mass of mandatory and advisory requirements concerning the safety of *all* products, we offer in the following pages a proven management systems approach for product safety, which may be modified to suit the individual nature of the product. Some products, though not as complex as others, nevertheless require some form of systematic control for their manufacture and marketing. Other products may be very sophisticated and require complex planning in the various phases from concept to customer. This chapter will provide management with policy guidelines and an effective program to achieve the ultimate degree of safety of the product, within reason.

The philosophy of product or systems safety has been approached by management in various ways. In some instances, as government, business/industry, and newspaper reports indicate, it appears to have been severely neglected.

A system is an orderly combination or arrangement of various parts or elements into a whole. It may be defined as a major system composed of many subsystems. As an example, an automobile may be considered a system. The various elements (engine, transmission, suspension, air conditioner, radio, etc.) may be considered subsystems of the automobile proper. Each of these items standing alone may be considered a system, but when they become parts of the whole they are considered subsystems. These same subsystems may be broken down further into subsystems within themselves (e.g., the engine contains a carburetor, an entity unto itself, which in turn is made up of other items).

In talking about a system in this chapter, we will be speaking of the total "thing." Its total makeup will be thought of as including every item associated with it, including its maintenance and operational manuals, training (if required), etc., since a weakness which may adversely affect the proper operation of the total thing, within its working environment, then becomes the proverbial nail that lost a kingdom.

CURRENT PRACTICES AND PROBLEMS

Some corporations have outstanding systems safety programs (notably segments of the aerospace industry) which were originally generated by the nature of their developments. Though systems in general have been used for a considerable period of time, the matter of *systems safety* was refined in the 1940s with the advent of nuclear weaponry. The potentially catastrophic effects of a serious nuclear accident could not be tolerated. Consequently, every effort was made to insure that no error would occur and that there would be sufficient back-up, fail-safe devices to forestall any unwanted reaction. The advent of the space and missile age in the early 1950s brought *formalized systems safety* to maturity. The U.S. space and ballistic missile programs, with the complexity of their hard-

ware and associated subsystems spawned every con-
ceivable means of safety controls. The risks of disastrous occurrences were immeasurable. With the advent
of these programs a new and comprehensive military
standard was created, "System Safety Program for
Systems, Associated Subsystems and Equipment:
Requirements For" MIL-STD-882. This standard
subsequently became and is still a contractual requirement for industry by the Department of Defense (DOD)
and is used for developments of complex and hazardous
programs. (The author hopes that DOD policy will
call for judicious use of this standard because of its
heavy, all-inclusive, blanket coverage.)

These space and missile programs to a great extent
developed the formalized systems safety efforts of
today, by acting as a catalyst to coordinate input of
systems safety techniques which originated throughout
the broad range of U.S. industry. Many major companies in various industries today use applications of
these principles and techniques in their product safety
programs. They may be scaled down to fit the design
and production of relatively simple and far less sophisticated items than aerospace or other inherently complex
products.

Some managers believe that the matter of safe
design and performance will take care of itself if the
engineers and designers are left alone to do their jobs,
without any additional safety input from the corporate
and consultative brainpower pool. This approach has
been shown to be relatively ineffective. Though certain
physical results may be achieved, the resultant lack
of coordination with other departments within the
organization creates voids, misunderstandings, and

confusion and may require subsequent rework or redesign of product. Here we have a general hit-or-miss situation, in which *safety* of the product or system often becomes a tagalong thing.

Some approaches vest the entire matter of product/ systems safety in one or another of the organization's departments (engineering, quality control, etc.), without sufficient broad—or even limited—guidelines. This may result in a disjointed situation wherein total coordination is not achieved with all responsible managers and specialists. The degree of risk may be less than in the previous approach, but is still present in varying degree.

Some say that the importance of product or systems safety is vastly overrated, and that if everyone is left alone to do his job, things will turn out all right. In the opinion of the author, this approach is the worst possible. By ignoring a logical method of internal planning, controlling, and coordination, as well as disregarding the lessons of the past, we again arrive at a hit-or-miss situation and the element of risk is greatly magnified.

Some believe that the nature of the product is so simple that no effort at all is required to consider its safety, only to subsequently discover that the material that went into a line of children's dresses may cause those dresses to go up in flames when they come in contact with a spark.

DEVELOPMENTS, RISKS, AND OBLIGATIONS

A change in some "traditional" ways of thinking is warranted by the general sophistication and complexity of many products today, coupled with the use of new and exotic materials which may react unfavorably under certain circumstances, the mass of statutory, mandatory, and advisory requirements concerning product safety, and the mounting numbers of legal negligence actions. Hardly a day or week passes when one does not read of an injury, a lawsuit, damage or recall of a product, based upon potential or alleged product defect. Quite recently, the discovery of a hazardous "trouble light"—hazardous because it could potentially electrocute its users—resulted in recall of the entire product line, possible negligence actions, considerable front-page coverage in the press, and even a special TV report by one of the major networks. A judge decided that the requirements of the Consumer Product Safety Act (whereby the company in question must advertise and otherwise notify its dealers, distributors, and so on, as well as the general public, of the hazardous nature of the product, and its recall) had already been met by the intensive national coverage in the news media.

This serves to emphasize one focal point. The potential risks of some current methods of design and manufacture can no longer be tolerated. These risks warrant as much intensive review as the periodic corporate financial reports which are submitted to all managers. The total impact of the National Electronic Injury Surveillance System (NEISS) concerning products involved in accidents, and the investigation into causative factors, hasn't yet been felt. A logical company approach in reviewing all factors bearing on and controlling those risks would be to create a systematized approach and a product safety committee.

This combination would assure the necessary technical and professional know-how for establishing a minimum safety baseline for the product. More about this later.

Industry has always had an inherent moral obligation to protect the consumer by creating products which are tolerably hazard-free. Generally, much of U.S. industry has endeavored to do this, and has been doing it for a long time. Unfortunately, a major segment of industry has not been seriously fulfilling its responsibilities, as witnessed by the toll of accidents attributable to product deficiencies. The full scope of the problem hasn't yet been calculated, but the current feedback of information into the government's data collection system should unfold the story.

THE NEED FOR A LOGICAL AND MANAGEABLE SOLUTION

As previously pointed out, the sweeping requirements of the Consumer Product Safety Act, the Occupational Safety and Health Act, and other statutes governing specific products and industries, as well as current standards, pose a major challenge to U.S. industry today. Since the CPSA and OSHA require constant updating of standards based on research and changing developments, the scope of the problem becomes manifold. The current focus in U.S. industry today is on compliance with OSHA, but in the opinion of the author the CPSA and its far-reaching provisions pose a broader and more potent challenge. The combination of these requirements, with the possibility of punitive action, means that hit-or-miss business methods will be neither fruitful nor economical in the long run.

Whereas OSHA specifies physical requirements for occupational safety and health, and CPSA prescribes requirements for protection of the consumer against unreasonable risks of injury associated with consumer products, neither of these statutes points out to any appreciable degree the added beneficial effects which may accrue to industry by strict compliance. The beneficial effects of accident prevention have been explored in some detail in previous chapters. Relative to product safety, the CPSA, properly pursued by government and backed up by industry's cooperation, should bring about the following:

1. *Safer consumer products.*
2. *Safety in marketing, advertising, and labeling,* as well as all related aspects of the safety of products.
3. *Determination of target products* for priority action as well as creation of new standards.

A methodical approach to compliance will bring still other byproduct or submerged benefits to the corporation:

1. *Economy* of product development and production, through a planned program of product hazard elimination.
2. *Coordination* through staffing action between all involved departments, and through brainstorming the problems in their earliest stages.
3. *Awareness* of the relationship between corporate managers, legal counsel, designers, and engineers, and of their individual impact on the final product.

4. *Financial savings* by eliminating, to a great degree, the need for return of products to the plant for rebuild or repair, excessive field modification or repair, or the requirement for overly large field staffs of engineers (or mechanics) for correction of problems which might have been corrected in the first place through proper design and construction.
5. *Marketing benefits* through elimination of adverse reactions from dissatisfied customers, dealers, and salesmen.
6. *Maintenance of the corporate image* through elimination of the need to withdraw products from the marketplace.
7. *A competitive edge* because of all of the foregoing.
8. *Provision of a strong and coordinated data base for use in legal actions over products.*

At this point we shall present a set of workable guidelines for small business. These will be followed by a broad program for management action in major enterprises.

A GUIDELINE FOR SMALL BUSINESS

In the matter of total product safety, whether covered by the CPSA or other statutes, the *minimum approach* which a small business might reasonably take can be established by asking certain basic questions:

1. What is the problem, or what do we want to produce?
2. What are the factors bearing on the problem?
3. What are the possible solutions?

4. What are the tests of the possible solutions?
5. What is the best solution?

Responses to these questions should lead step by step to a reasonable decision:

1. Determine the best and safest product that can be produced, within available resources and conforming to current statutes, to meet a customer need.
2. Provide the designers and engineers with the requirements, based on past experience and present limitations, and backed up by information from trade associations and insurance companies. These should be accompanied with instructions that the designers/engineers conform to regulatory requirements. (Steps one and two may be considered the "concept" phase.)
3. Narrow the possibilities to the most desirable solutions which the designers/engineers may create, within the guidelines of existing laws, standards, and other requirements.
4. The solutions must be subjected to tests, analysis, and trade-offs to determine the best and safest product. (Much of the information may be secured from subcontractors and suppliers, concerning their own products or materials.)
5. The result will be the best and safest product that can be produced, within reason. (The immediate solution.)

(Here is the end of the design and engineering phase and the groundwork for production).

Complete documentation should be kept of all planning, actions, tests, quality control, and other procedures taken during the manufacturing process, to insure safety of the product. Should any customer complaints subsequently be received, concerning safety of the product, records of corrective actions and changes should be included in the file. These records may be continuously expanded and kept as a permanent file, not only to assist legal counsel in the event of a liability suit against the company, but to use as a technical information source for future product design and manufacture.

A BROAD PROGRAM FOR MANAGEMENT ACTION

The degree of safety achieved in any product or complex system is directly related to and dependent upon management emphasis. To reasonably achieve the goal of product safety demands that safety considerations be incorporated into all phases of the product's life cycle, from concept through engineering development (and design) to production and consumer/ operational use. In order to provide management with a wide-ranging overview of the program needed to establish a heavy degree of safety coverage and control during the product or system life cycle, use of the following broad guidelines is strongly recommended. (The guidelines may be expanded, contracted, or reasonably tailored to fit specific products or developments and to establish company policy.)

Management policy should be written out, and should clearly state the total objectives and requirements before the product is designed, or at the inception of a major system, rather than permit safety considerations to be applied as an afterthought, after product design.

To do the latter usually means reworking, redesigning, and modifying the item, with resultant dislocation and increased costs.

When the scope or magnitude of any program does not warrant a comprehensive program, management may logically specify the minimum acceptable safety standards to be met.

Prudent management should strive for the optimum degree of safety which can be realized within the boundaries of legislative requirements, existing standards, operational effectiveness, time, and cost. The degree of emphasis given to the programmed results which are to be achieved will consequently protect the consumer, and thus insure against legal liability claims which might follow should a consumer be exposed to accident or injury because of bad design.

Top management should establish a coordinated activity (say a product safety committee composed of key management and technical representatives from all major departments, safety, legal counsel, and other related specialized fields) within the organization, *charged with providing the baseline to insure that:*

1. Safety consistent within economic limitations and operational effectiveness is designed into the product or system.
2. Hazards associated with the product and its subsystems, parts, and materials, are identified, evaluated, eliminated, or controlled to an acceptable and tolerable level.
3. Labeling and markings clearly define the existing and potential hazards, as well as give cautionary instructions.

4. Control over hazards which cannot be eliminated is established, and incorporated into operational/instructional/maintenance publications to protect the user, equipment, general public, and environment.

5. Minimum risk is involved in acceptance and use of new materials, as well as new production and testing techniques.

6. After-the-fact and retrofit (return for modification, repair, or rebuild) actions required to improve safety are minimized through timely addressing of the safety requirements during initial concept, and subsequent design, engineering development, testing, and production.

7. Thorough analysis is accomplished during all phases of development, dependent upon the complexity of the product.

8. Historical data generated by similar developments is maintained and applied toward achieving the final and best product, as well as providing a data bank for future developments, and legal support.

9. Accident information secured from NEISS, the National Safety Council, insurance carriers, trade associations, technical societies, legal liability actions, consumer complaints, distributors, etc., is maintained, researched, and thoroughly considered in design of the new product. This information would also be incorporated into the data bank.

10. Mandatory and voluntary safety standards are systematically followed.

The combination of the foregoing with suitable records and documentation of actions which the organization has taken to eliminate product hazards will not only protect the customer, but will give the company a reasonable base with which to provide a better product and defend itself against possible future liability suits.

MANAGEABLE PHASES OF THE PROGRAM AND RELATED ACTIONS

Putting the entire process of achieving the ultimate in safety in our products (within the stated guidelines) into manageable perspective, *requires that coordinated safety considerations be monitored* in the following phases of the life-cycle (dependent upon the complexity of the product).

Concept

At the outset, and *before engineering design and development,* when it is determined that a certain major product will be manufactured, the product safety committee will determine and prescribe the broad guidelines and parameters that the engineers will use in designing the product. To help provide those guidelines (or baselines), the committee may lean heavily on historical material, or the safety and legal data bank of information maintained by the company, trade associations, etc. (Note: The term "data bank" does not necessarily mean a computer. Such a bank could be maintained in a coordinated file.)

At this point, management establishes the overall

policy and principles to be followed so that the product performs in its intended operational environment or working situation. To insure that safety characteristics are adequately addressed, the coordinating group would: (1) evaluate technical approaches to safety design features; (2) identify possible safety interface problems, e.g., operator/machine/working environment, system/subsystems/materials, packaging/shipping, instructional materials/training (of operators, salesmen, dealers), weather and environmental conditions; and (3) highlight areas of safety consideration, such as systems limitations and capabilities, risks, human factor limitations, fatigue, environmental impact, production considerations, materials, marketing factors, and testing techniques. Finally, it would define areas requiring further hazard investigation based on historical data and on tests which may be required to eliminate or minimize such hazards.

Design & Engineering Development

Using the above broad guidelines, and conforming to applicable mandatory and advisory standards, the designers/engineers would then proceed to design the product, and consequently design-out potential hazards. These specialists, having been made aware of the baseline of safety characteristics to be followed, would be in a stronger position to insure against costs, efforts, time loss, and adverse customer reaction in correcting hazards which should have been eliminated in the first place.

Coordination should continue to be made with key managers so that they may not only provide their own input, but also be made aware of what is coming in the way of new testing, production, materials, or personnel or training requirements. These managers may then preplan (if only in their minds) their new or additional requirements. No manager likes to be surprised, if he can help it.

During this phase the following would be accomplished:

1. *Analyses to determine the product's technical safety specifications, performance requirements, and operational safety characteristics.*

2. *A preliminary hazard analysis of the product in its operational environment,* that is, in the place and under the various circumstances in which it is designed to be used.

3. *Product and subsystem/equipment/materials/ sub-contract-items interface studies* to insure compatibility and nondegradation of safety. To further prevent introducing a weak link, management (and its technical personnel) should insure that subcontract material/items be produced or tested under current standards and rigid specifications, so that they, in turn, conform to desired standards.

4. *Trade-off approaches* to identify inherent risks and required controls.

5. *Analyses of operator/user physiological and psychological stresses, and related human factors,* to avoid design which might generate errors leading to accidents.

6. *Incorporation of fail-safe features/devices,* within reasonable cost and operational considerations, wherever a failure might result in critical injury or fatality.

7. *Inclusion of qualitative and quantitative testing requirements* in specifications and proposed test plans.

8. *Product and subsystem hazards and operational analysis* (e.g., failure modes and effects analysis, or more sophisticated analyses such as fault-free), depending upon the degree of sophistication of the product.

9. *Testing* of materials, subassemblies, contracted items/materials, prototypes. If there are insufficient or inadequate in-house facilities for this testing, then use should be made of certified testing laboratories.

10. *Identification and specification of cautions, warnings, labeling, and related controls* for the product and consumer, and inclusion in operational, training (if required), and maintenance publications and promotional materials.

11. *Identification and specification of protective equipment and clothing* requirements.

12. *Establishment of marketing and dealer/customer training requirements* for complex items, providing thorough safety coverage.

Production/Manufacturing

The design and engineering process is finalized after a closely reviewed de-bugging period. Production manufacturing begins, and quality control (or product assurance) testing of in-house and control of contracted items/materials insures that the product is manufactured to desired specifications. During this testing, undetermined hazards may be identified, either in the product or manufacturing techniques, and can be corrected after close coordination with development engineers and manufacturing managers, as well as subcontractors.

Finally manufacturing processes should be closely monitored to eliminate product deficiencies which might be created on the assembly line through either mechanical, physical, or human error (fatigue, etc.).

Customer Use/Marketing

During this phase, when the product is in the hands of the dealers and the customers, the following remains to be done:

1. *Determine whether design, operating, maintenance, and emergency procedures have proven adequate,* based on user/dealer experience.
2. *Evaluate subsequent design changes and modifications to insure that safety is not degraded.*
3. *Maintain continual review of operational, maintenance, and marketing publications* to insure that safety requirements, procedures, and precautions are not only adequate, but are suitably highlighted to protect the consumer.
4. *Analyze equipment accidents or failures* which have caused, or could cause, an unsafe condition, and correct them.
5. *Collect and analyze product deficiency reports* received from customers and dealers.
6. *Provide identification, using procedures, precautions, and disposal requirements for dangerous*

> *materials* or components (toxics, explosive devices, radioactive materials, etc.).
>
> 7. *Maintain historical data file (or bank)* from the above reports. This, in turn, will provide material for the product safety committee, legal counsel, and all other managers, for future efforts and research.

SUMMARY

Completing the complex phases outlined above may appear to be a formidable task, but it must be remembered that depending upon the complexity of the product or its related materials, it can be expanded or reduced at will. The full application of the total product/systems program should be judiciously used. If your corporation is involved in the manufacture or development of a complex item of hazardous military equipment, it is most probable that the foregoing demands will be a contractual requirement. If your product line is consumer oriented, yet is inherently complex or hazardous, the principles also apply. One thing for sure, monitoring and coordination for this activity requires a highly trained product safety manager. To settle for anything less in the interests of "expediency" is to place management, and the company, in the same risk category as the product.

Whatever the case, using these principles and procedures to achieve the optimum degree of safety within the limits of customer use, effectiveness, time, and cost, will protect the consumer, assist in compliance with current product safety requirements (legislative and advisory), eliminate the need for dealer or factory repair, rebuild, or recall of the product, eliminate most of the causes of consumer legal actions, protect, to a great degree, against such legal actions, and thrust the company's image above the competition's. The use of the principles will provide for timely identification of hazards, as well as initiation of those actions necessary to prevent or control such hazards, and will provide a reasonable base from which to specify, predict, and evaluate the safety of the end product.

Demands on future products and materials will be complex. Complexity in requirements, however, does not necessarily mean complexity of design. While technology has taken giant strides, the very nature and physical makeup of the human beings who will operate equipment and use products remains relatively static. We must always bear in mind the human factors required of the consumer who will use the product.

12

The National Business Council for Consumer Affairs (NBCCA)

A recent development in the matter of consumer product safety came with the publication in 1974 of *Action Guidelines—National Business Council for Consumer Affairs*[1] as a public service by Montgomery Ward & Co. This publication is a compilation of significant reports issued by the National Business Council for Consumer Affairs (NBCCA). The NBCCA was established by the U.S. Secretary of Commerce through a presidential directive of August 5, 1971. The directive pointed out that ". . . neither the government nor the consumer movement can alone solve these problems, but that we must also rely upon the traditional goodwill and sound practices of the business community."

---

[1] *Action Guidelines—National Business Council for Consumer Affairs*, Montgomery Ward & Co., Consumer Information Services, 20–N, 535 W. Chicago Ave., Chicago, Ill. 60607.

Organization and Representation

Various subcouncils of the NBCCA Advisory Committee representing the business community were established, and provided their individual reports and recommendations. Though the reports contain the results of studies by the Advisory Committee and its subcouncils, they do not necessarily represent the views of the Department of Commerce or of any other federal agency.

It is meaningful that 116 leaders in the production and distribution of consumer goods developed these action guidelines, and that they are already being followed by their companies. The subjects addressed in the final reports are:

1. Advertising and promotion.
2. Packaging and labeling.
3. Financing and the American consumer.
4. Product performance and servicing.
5. Safety in the marketplace.
6. Complaints and remedies.
7. Advertising substantiation.

SUBCOUNCIL REPORT, "SAFETY IN THE MARKETPLACE"

In 1973 the NBCCA's subcouncil on product safety prepared a report entitled "Safety in the Marketplace: A Program for the Improvement of Consumer Product Safety." In its covering letter submitting the report to the Secretary of Commerce, the subcouncil stated:

> Virtually all of the recent studies and dialogue on consumer product safety have centered upon the legislative issues relating to the recently enacted Consumer Product Safety

Act. The Sub-Council recognizes the importance of this Federal action and its report focuses on private sector activities important to the operation of the new Product Safety Commission, as well as those which should be strengthened regardless of the form and magnitude of the Federal role in this area. Thus, the report contains analysis, conclusions, and recommendations relating to manufacturers, retailers, voluntary standards organizations, product testing laboratories, and trade associations, as well as government and other organizations in a position to contribute to improved product safety.

The issues covered in the report are complex and often technical, but we believe they should be understood by all concerned with product safety. We therefore recommend that the report be given wide distribution within the government and in the private sector.

The summary and recommendations of the sub-council concerning the manufacturer's role, as published in *Action Guidelines — National Business Council for Consumer Affairs*, are as follows:

I. BACKGROUND

The National Commission on Product Safety has estimated that 20 million injuries occur each year in and about the household. Individuals, products, and their environment, are all involved in these accidents, and the contributing factors are varied and complex. As a result, dramatic reductions in injury rates cannot be expected from any single action. However, incremental improvements over time are possible, and they constitute a worthwhile national goal. Many institutions are in a position to contribute to this end, including corporations, trade associations, technical organizations, and government agencies.

II. THE MANUFACTURER'S ROLE

Manufacturers have the responsibility to assure that the safety of the product is taken into account throughout the activities involved in its design, production, and distribution. This is best accomplished by a comprehensive systems approach to product safety, which includes every step from the creation of a product design to the ultimate use of the product by the customer. To assure that this system is effective:

Recommendation 1
Corporations which manufacture consumer products should maintain and enforce written safety policies, standards and procedures on all pertinent corporate activities. These documents should be regularly reviewed and updated.

These procedures should include the establishment of a broadly based internal safety review council and should provide guidelines for product design, testing, quality assurance, and product identification procedure.

Insurance companies can assist manufacturers in creating and maintaining the policies and procedures required for adequate product safety. This would be especially meaningful for those manufacturers with limited technical staffs. Therefore:

Recommendation 2
Insurance companies offering product liability coverage should develop techniques to assist their corporate clients in establishing effective product safety practices.

This independent and confidential review of clients' procedures by insurance companies would be a helpful check on their adequacy from a safety point of view and would complement advice and assistance from other sources, such as independent laboratories, safety consultants, universities and government agencies.

Manufacturers also can contribute to the safe use of their products by emphasizing safety in the wide range of information services provided to customers. Because advertising (including packaging and display materials) is often the most pervasive corporate contact with the consumer:

Recommendation 3
All advertisers should review the content of their messages for safety implications.

The report contains sections on the procedures to use in every step of production and marketing, from initial design to final marketing and service. We recommend that it be studied in depth. (Details on how to secure this information are given in appendix B.) The report provides factual data to complement and supplement statutory requirements of the Consumer Product Safety Act. It further supports existing mandatory and advisory standards concerning product safety. Finally, it offers astute recommendations for development of new standards, based on accident information and related data generated by the Consumer Product Safety Commission's National Electronic Injury Surveillance System.

13

The Third Industrial Revolution

INDUSTRIAL REVOLUTION . . . The period of economic transition from hand industry to the application of power-driven machinery which began in England in the second half of the 18th century.

Funk & Wagnalls New College Standard Dictionary

The Industrial Revolution that we learned about in school was the transformation of British industry during the latter part of the 18th century from a handicraft to a machinery basis. A series of mechanical discoveries brought this about. The first was the discovery that iron could be smelted with coke. This discovery changed the district of the north of England, where coal and iron lie side by side, into an industrial center. A few years later, Watt's discoveries applied steam to the working of machinery, and Arkwright changed spinning and weaving from a hand to a machinery operation. All of Britain's European rivals, which were handicapped by antiquated methods, as well as by internal strife, were then hopelessly outdistanced. In a single generation British industry supplanted that of the other nations of the world. Factories mushroomed, and the population of the north quadrupled in 30 years. To some extent, this is about where the economic impact of industrial history, as we were formally taught it in school, stops. I therefore choose to call this era "The First Industrial Revolution."

Subsequently came the emergence of other industrial powers, and another occurrence came about which was to change history: the harnessing of electricity and the development of electronic technology. I choose to call this "The Second Industrial Revolution," or the "electronic" revolution. New industries were created and expanded to the point where their workings stretched the imagination. The discoveries of Alessandro Volta and the electrical spark generated by Benjamin Franklin betokened the emergence of new eras of power supply and control. The metamorphosis from wire communications to radio, and to the electronic marvels of today, opened up yet another chapter in industrial growth and expansion.

Today, memory machines and automatic data-processing equipment, as well as tape-run industrial machines, are considered standard equipment. With them, new fields of industrial conquest opened up, and the sights of inventors were raised higher than ever before. And then other massive discoveries came about: *the harnessing of the forces of nature.* I call this "The Third Industrial Revolution," or the physical revolution.

Waterpower was controlled to a degree never imagined by the Romans. Hydroelectric power complexes sprang up. The process of desalination, which converts sea water into fresh water for drinking and agriculture, opens up the capability of rejuvenating the arid areas of the world. Thermodynamics and solar energy, married to electronics, are opening up new areas where the common energy sources of today will be pushed into the background tomorrow. The use of energy from the earth has opened up a new source of power and heat. Iceland, for example, is drawing a considerable amount of power from the banked furnaces of old volcanoes, and our own northwestern states are harnessing steam from the earth's underground kettles. The discovery and creation of nuclear energy has released a genie which has not yet flexed its muscles, despite the fact that it has already changed history. The transformation of the earth's bounty into changed physical sources (e.g., shale deposits into oil) has begun. We are only starting to harvest and mine the sea, to extract its unlimited sources of protein, as well as manganese and unknown quantities of many other metals.

Though we've come a long way, the road stretches far ahead of us. The above accomplishments have come about primarily in this century, with several nurtured in just the last three generations. In our lifetime we will see new discoveries which will make present developments obsolete, and all of the advances already made open up new and broad areas of hazards, and call for dynamic measures for control of those hazards. Management must therefore take a close look at itself and ask, "Do we have the safety professionalism and capability in-house to keep up?"

Appendix A

Sources of Information

Federal Agencies

U.S. Department of Labor

Occupational Safety and Health Administration (OSHA)
U.S. Department of Labor Building
14th Street and Constitution Avenue, N.W.
Washington, D.C. 20210
Bureau of Labor Standards
Washington, D.C. 20210
Bureau of Labor Statistics
441 G Street, N.W.
Washington, D.C. 20210

U.S. Department of Health, Education, and Welfare

National Institute of Occupational Safety and Health (NIOSH)
5600 Fishers Lane
Rockville, Md. 20852
Consumer Product Safety Commission
Washington, D.C. 20207

122

Bureau of Product Safety
  5401 Westbard Avenue
  Bethesda, Md. 20016
Public Health Service
  Washington, D.C. 20203

U.S. Department of the Interior
Bureau of Mines
  Washington, D.C. 20240

U.S. Department of Transportation
  800 Independence Avenue, S.W.
  Washington, D.C. 20590
Federal Aviation Administration
  800 Independence Avenue, S.W.
  Washington, D.C. 20590

U.S. Atomic Energy Commission
  Washington, D.C. 20545

U.S. Department of Agriculture
Entomology Research Division
Pesticide Research Branch
  Agriculture Research Center
  Beltsville, Md. 20705

Superintendent of Documents
Government Printing Office
  Washington, D.C. 20402

Occupational Safety and Health Administration (OSHA)

Regional and Area Offices

Region I: Connecticut, Maine, Massachusetts, New
           Hampshire, Rhode Island, Vermont
  18 Oliver Street
  Boston, Mass. 02110

Area Offices:

450 Main Street — Rm. 617
Hartford, Conn. 06103

Custom House Building
State Street
Boston, Mass. 02109

55 Pleasant Street — Rm. 426
Concord, N.H. 03301

436 Dwight Street — Rm. 501
Springfield, Mass. 01103

Region II: New York, New Jersey, Puerto Rico,
Virgin Islands, Canal Zone
1515 Broadway
New York, N.Y. 10036

Area Offices:

970 Broad Street — Rm. 1435C
Newark, N.J. 07102

370 Old Country Road
Garden City, L.I., N.Y. 11530

90 Church Street — Rm. 1405
New York, N.Y. 10007

700 East Water Street — Rm. 203
Syracuse, N.Y. 13210

605 Condado Avenue — Rm. 328
Santurce, P.R. 00907

Region III: Delaware, District of Columbia, Mary-
land, Pennsylvania, Virginia, West
Virginia
Gateway Center — Suite 15220
3535 Market Street
Philadelphia, Pa. 19104

Area Offices:

31 Hopkins Plaza — Rm. 1110
Baltimore, Md. 21201

Jonnet Building — Rm. 802
4099 William Penn Highway
Monroeville, Pa. 15146

600 Arch Street — Suite 4456
Philadelphia, Pa. 19106

Federal Building
400 N. 8th Street — Rm. 8018
P. O. Box 10186
Richmond, Va. 23240

700 Virginia Street — Suite 1726
Charleston, W.Va. 25301

Region IV: Alabama, Florida, Georgia, Kentucky, Mississippi, North Carolina, South Carolina, Tennessee

1375 Peachtree Street, N.E. — Suite 587
Atlanta, Ga. 30309

Area Offices:

2047 Canyon Road, Todd Mall
Birmingham, Ala. 35216

118 North Royal Street — Rm. 600
Mobile, Ala. 36602

3200 E. Oakland Park Blvd. — Rm. 204
Fort Lauderdale, Fla. 33308

La Vista Perimeter Park — Suite 33
Building 10
Tucker, Ga. 30384

2720 Riverside Drive
Macon, Ga. 31204

6605 Abercorn Street — Suite 204
Savannah, Ga. 31405

600 Federal Place — Rm. 554-E
Louisville, Ky. 40202

2809 Art Museum Drive — Suite 4
Jacksonville, Fla. 32207

310 New Bern Avenue — Rm. 378
Raleigh, N.C. 27601

1710 Gervais Street — Rm. 205
Columbia, S.C. 29201

5760 I-55 No. Frontage Rd. East
Jackson, Miss. 39200

1600 Hayes Street — Suite 302
Nashville, Tenn. 37203

Region V: Illinois, Indiana, Michigan, Minnesota,
Ohio, Wisconsin
230 So. Dearborn — 38th Floor
Chicago, Ill. 60604

Area Offices:
230 So. Dearborn — 10th Floor
Chicago, Ill. 60604

46 East Ohio Street — Rm. 423
Indianapolis, Ind. 46204

220 Bagley Avenue — Rm. 626
Detroit, Mich. 48226

110 South Fourth Street — Rm. 437
Minneapolis, Minn. 55401

550 Main Street — Rm. 4028
Cincinnati, Ohio 45202

1240 East Ninth Street — Rm. 847
Cleveland, Ohio 44199

360 So. Third Street—Rm. 109
Columbus, Ohio 43215

234 N. Summit Street—Rm. 734
Toledo, Ohio 43604

633 W. Wisconsin Avenue—Rm. 400
Milwaukee, Wis. 53203

Region VI: Arkansas, Louisiana, New Mexico,
Oklahoma, Texas
1512 Commerce Street—7th Floor
Dallas, Tex. 75201

Area Offices:

103 East 7th Street—Rm. 526
Little Rock, Ark. 72201

546 Carondelet Street—Rm. 202
New Orleans, La. 70130

1015 Jackson Keller Road—Rm. 122
San Antonio, Tex. 78213

420 South Boulder—Rm. 514
Tulsa, Okla. 74103

1412 Main Street—Suite 1820
Dallas, Tex. 75202

2320 LaBranch Street—Rm. 2118
Houston, Tex. 77004

1205 Texas Avenue—Rm. 421
Lubbock, Tex. 79401

421 Gold Avenue S.W.—Rm. 302
P.O. Box 1428
Albuquerque, N.M. 87103

Region VII: Iowa, Kansas, Missouri, Nebraska
911 Walnut Street—Rm. 3000
Kansas City, Mo. 64106

Area Offices:

1627 Main Street—Rm. 1100
Kansas City, Mo. 64108

210 North 12th Boulevard—Rm. 554
St. Louis, Mo. 63101

16th and Harney Street
City National Bank Building—Rm. 803
Omaha, Nebr. 68102

221 South Broadway Street—Suite 312
Wichita, Kans. 67202

Region VIII: Colorado, Montana, North Dakota, South Dakota, Utah, Wyoming

1961 Stout Street—Rm. 15010
Denver, Colo. 80202

Area Offices:

8527 W. Colfax Avenue
Lakewood, Colo. 80215

2812 1st Avenue North—Suite 525
Billings, Mont. 59101

455 East 4th South—Suite 309
Salt Lake City, Utah 84111

Region IX: Arizona, California, Hawaii, Nevada, Guam, American Samoa, Trust Territory of the Pacific Islands

450 Golden Gate Avenue—Rm. 9470
P.O. Box 36017
San Francisco, Calif. 94102

Area Offices:

2721 North Central Avenue—Suite 318
Phoenix, Ariz. 85004

19 Pine Avenue—Rm. 401
Long Beach, Calif. 90802

100 McAllister Street—Rm. 1706
San Francisco, Calif. 94102
333 Queen Street—Suite 505
Honolulu, Hawaii 96813
1100 East William Street—Suite 222
Carson City, Nev. 89701

Region X: Alaska, Idaho, Oregon, Washington
506 Second Avenue—Rm. 1808
Seattle, Wash. 98104

Area Offices:
605 West 4th Avenue—Rm. 227
Anchorage, Alaska 99501
921 S.W. Washington Street—Rm. 526
Portland, Oreg. 97205
121 107th Avenue, N.E.—Rm. 110
Bellevue, Wash. 98004
228 Idaho Building
Boise, Idaho 83702

Bureau of Labor Statistics

Regional Offices

Region I: Connecticut, Maine, Massachusetts, New
Hampshire, Rhode Island, Vermont
1603-A Federal Office Building
Boston, Mass. 02203

Region II: New Jersey, New York, Puerto Rico, Virgin Islands
1515 Broadway
New York, N.Y. 10036

Region III: Delaware, District of Columbia, Maryland, Pennsylvania, Virginia, West Virginia

P.O. Box 13309

Philadelphia, Pa. 19101

Region IV: Alabama, Florida, Georgia, Kentucky, Mississippi, North Carolina, South Carolina, Tennessee

1371 Peachtree Street, N.E.

Atlanta, Ga. 30309

Region V: Illinois, Indiana, Michigan, Minnesota, Ohio, Wisconsin

230 So. Dearborn—9th Floor

Chicago, Ill. 60604

Region VI: Arkansas, Louisiana, New Mexico, Oklahoma, Texas

555 Griffin Square Building—2nd Floor

Dallas, Tex. 75202

Regions VII and VIII: Colorado, Iowa, Kansas, Missouri, Montana, Nebraska, North Dakota, South Dakota, Utah, Wyoming

Federal Office Building

911 Walnut Street

Kansas City, Mo. 64106

Regions IX and X: Alaska, Arizona, California, Hawaii, Idaho, Nevada, Oregon, Washington

450 Golden Gate Avenue

Box 36017

San Francisco, Calif. 94102

Reference Organizations and Originators of Standards

Note: An asterisk (*) denotes organizations originating standards referenced in the blanket coverage of 29 CFR Part 1910.6 of the Occupational Safety and Health Act of 1970. A dagger (†) denotes organizations originating standards referenced in 41 CFR Part 50–204, *Safety and Health Standards for Federal Supply Contracts.* Since this code of standards is repeatedly referenced in the Occupational Safety and Health Act of 1970, and the standards apply to that act as well, the originating organizations are included in this appendix to eliminate confusion on the part of the reader of the Act of 1970 and provide a ready reference. A double dagger (‡) denotes organizations originating standards for both the act and the code.

American Conference of Governmental Industrial Hygienists (ACGIH)‡
1014 Broadway
Cincinnati, Ohio 45202

American Industrial Hygiene Association (AIHA)
25711 Southfield Road
Southfield, Mich. 48075

American National Standards Institute (ANSI)‡
1430 Broadway
New York, N.Y. 10018

American Petroleum Institute*
1801 K Street, N.W.
Washington, D.C. 20006

American Plywood Association*
1119 A Street
Tacoma, Wash. 98401

American Society of Agricultural Engineers (ASAE)*
2950 Niles Road
St. Joseph, Mo. 49085

American Society of Heating, Refrigeration and Air
Conditioning Engineers, Inc.*
345 East 47th Street
New York, N.Y. 10017

American Society of Mechanical Engineers, Inc.
(ASME)‡
345 East 47th Street
New York, N.Y. 10017

American Society of Safety Engineers (ASSE)
850 Busse Highway
Park Ridge, Ill. 60068

American Society for Testing and Materials‡
1916 Race Street
Philadelphia, Pa. 19103

American Welding Society‡
44 Madison Avenue
New York, N.Y. 10022

Compressed Gas Association‡
500 Fifth Avenue
New York, N.Y. 10036

Crane Manufacturers Association of America, Inc.*
1 Thomas Circle, N.W.
Washington, D.C. 20005

Factory Mutual Engineering Corp.*
P.O. Box 688
Norwood, Mass. 02062

Institute of Makers of Explosives*
420 Lexington Avenue
New York, N.Y. 10017

National Association of Plumbing and Mechanical Officials*
5032 Alhambra Avenue
Los Angeles, Calif. 90032

National Board of Boiler and Pressure Vessel Inspectors*
1155 North High Street
Columbus, Ohio 43201

National Committee on Uniform Traffic Laws and Ordinances†
525 School Street, S.W.
Washington, D.C. 20024

National Fire Protection Association (NFPA)‡
60 Batterymarch Street,
Boston, Mass. 02110

National Plant Food Institute*
1700 K Street
Washington, D.C. 20006

National Safety Council
425 North Michigan Avenue
Chicago, Ill. 60611

Power Crane and Shovel Association*
111 East Wisconsin Avenue
Milwaukee, Wis. 53202

Rubber Manufacturers Association*
345 East 47th Street
New York, N.Y. 10017

Society of Automotive Engineers (SAE)*
485 Lexington Avenue
New York, N.Y. 10017

Underwriters Laboratories, Inc.*
207 East Ohio Street
Chicago, Ill. 60611

Appendix B

Reference Publications

National Institute of Occupational Safety and Health (NIOSH)

Toxic Substances List, 1974 edition, HEW pub. No. (NIOSH)74–134: A single free copy may be obtained from the U.S. Department of Health, Education, and Welfare, Public Health Service, Center for Disease Control, NIOSH, Rockville, Md. 20852

Occupational Safety and Health Administration (OSHA)

Free Publications*

"Farm Employer and the Occupational Safety and Health Act of 1970, The" (OSHA 2010). A folder on the agricultural safety standards, as well as the rights and obligations of farmers under the OSHA act.

"Guidelines for Setting up Job Safety and Health Programs" (OSHA 2070). A booklet providing OSHA guidelines to assist employers in developing and implementing safety and health programs.

---

* Free copies of these publications are available from the OSHA office nearest you (see listing in Appendix A, for address).

"How States Plan for Job Safety and Health" (OSHA 2050). This pamphlet deals with state occupational safety and health programs; the content of the state plans, their criteria, the indices of their effectiveness, state standards, and the method of state enforcement.

"Inspection!" (OSHA 2026). Describes procedures followed by an OSHA compliance officer inspecting an establishment.

"Material Safety Data Sheet—Requirements For Reporting Hazardous Materials, Safety and Health Regulations for Ship Repairing, Shipbuilding, Shipbreaking."

"Noise—The Environmental Problem—A Guide to OSHA Standards" (OSHA 2067).

"Occupational Safety and Health Act of 1970—PL 91–596" (OSHA 2001). The full text of the Williams-Steiger Occupational Safety and Health Act of 1970.

"OSHA Fact Sheet for Small Businesses on Obtaining Compliance Loans" (OSHA 2005). A flyer outlining the procedures through which small business establishments can obtain OSHA assistance in applying for Small Business Administration loans to aid them in meeting OSHA standards.

"OSHA's Recordkeeping Requirements" (OSHA 2028). An interview with Thomas J. McArdle, the Assistant Commissioner, Bureau of Labor Statistics, who discusses the OSHA recordkeeping system and why it was needed.

"Recordkeeping Requirements Under the Occupational Safety and Health Act of 1970" (rev. 1975). A booklet containing employers' (including farm employers') responsibilities and new recordkeeping forms which must be used to record work-related injuries and illnesses which occur on or after January 1, 1975.

"Safe Use of Anhydrous Ammonia, The" (OSHA 2011). A folder describing the injury potential of anhydrous ammonia fertilizer, including precautions to take to prevent accidents, and what to do if an accident occurs while using this product.

"Safety and Health Protection On the Job" (OSHA 2003). Display poster stating purpose and scope of the OSHA act of 1970.

"Safety and Health Standards for Agriculture" (OSHA 2009). A booklet containing the complete text of the safety standards applicable to agriculture. Reprinted from the *Federal Register*, May 29, 1971.

"Selected Publications of the Occupational Safety and Health Administration" (OSHA 2019).

"Setting New Standards for Job Safety and Health" (OSHA 2027). This folder outlines the procedures for developing new safety standards and updating old ones.

"Target Health Hazards, The" (OSHA 2051). A booklet containing facts about five hazardous workplace substances which comprise OSHA's Target Health Hazards Program. The substances are asbestos, silica, lead, cotton dust, and carbon monoxide.

"Target Industries: A Profile of Five Hazardous Occupations" (OSHA 2034). A 24-page booklet documenting the safety problems of the five OSHA target industries. They are longshoring, roofing and sheet metal, meat and meat products, mobile homes and miscellaneous transportation equipment, and wood and wood products.

"Training Requirements of the Occupational Safety and Health Standards" (OSHA 2082). A booklet which is designed to help employers and employees identify the standards that relate to training.

"What Every Employer Needs to Know About OSHA Recordkeeping" (Report 412—Revised, U.S. Department of Labor, Bureau of Labor Statistics).

Publications for Sale*

"Handy Reference Guide: The Williams-Steiger Occupational Safety and Health Act of 1970, A" (OSHA 2004). Pocket-sized, 26-page guide to the act, describing its coverage, purpose, penalties, and other items every employer should know (20¢).

"Safety and Health Standards for Construction" (OSHA 2061). A 32-page booklet containing copies of the rules and regulations pertaining to construction (Title 29 CFR Part 1926) which were published in the *Federal Register* between April 18, 1971 and May 31, 1972 (25¢).

"Safety and Health Standards for General Industry" (OSHA 2060). A 24-page booklet containing the rules and regulations pertaining to general industry (Title 29 CFR Part 1910) which were published in the *Fed-*

---

* Copies of these publications are available from the Government Printing Office (GPO) at the prices given in parentheses. Make checks or money orders payable to the Superintendent of Documents, Government Printing Office, Washington, D.C. 20402. (Single copies may be obtained from OSHA Regional Offices, as well.)

eral Register between May 30, 1971 and May 31, 1972 (20¢).

"Volume 37, No. 75, Part II of the *Federal Register* of April 17, 1971." Contains the initial safety and health regulations for construction (Title 29 CFR Part 1926) under the Williams-Steiger Act (20¢).

"Volume 37, No. 105, Part II of the *Federal Register* of May 29, 1971." Contains the initial promulgation of general industry standards (Title 29 CFR Part 1910) under the Williams-Steiger Act (20¢).

"Volume 37, No. 202, Part II of the *Federal Register* of Oct. 18, 1972" — (Title 29 CFR Part 1910 update). Since the CFR will be constantly updated, check with your local OSHA Regional Office for current issues (20¢).

"Volume 37, No. 203, Part II of the *Federal Register* of Oct. 19, 1972." Safety and Health Regulations for Maritime Employment (rev.) (20¢).

"Volume 37, No. 243, Part II of the *Federal Register* of Dec. 16, 1972." Safety and Health Regulations for Construction (Title 29 CFR Part 1926 update) (20¢).

Available from OSHA National Headquarters, Washington, D.C. only:

"Employee Rights & Responsibilities — Under the Williams-Steiger Occupational Safety & Health Act of 1970" (OSHA 2017). This is a teaching guide which is part of a training package covering the salient facets of the act as well as the Occupational Safety and Health Administration and the program to improve injury and illness experience. (Consult OSHA Hq for price.)

"Employer Rights & Responsibilities—Under the Williams-Steiger Occupational Safety & Health Act of 1970" (OSHA 2009). Similar to above but geared to employers. (Consult OSHA Hq for price.)

Publications available from GPO only:

Code of Federal Regulations: Title 29 CFR Labor— Part 900 to End. A bound volume of the OSHA regulations as published in the *Federal Register.* Be sure to get the most current issue ($4.00 per copy).

Compliance Operations Manual (OSHA 2006). Contains mandatory guidelines for OSHA regional and area personnel to follow in implementing the OSHA act of 1970, including procedures for processing contested cases with the Review Commission, and information regarding citations, proposed penalties, and other OSHA activities. This document is considered a "must" since it provides a clear idea of what the OSHA representative will look for, the forms for doing his job, and an insight into the actions that will follow, in a vast number of situations. One of the best ways to prepare is to know precisely what the inspector is looking for ($2.00).

"Construction Safety & Health Training Course" (OSHA 2044). This is an instructor's guide for a 30-hour course, intended to provide general instructions concerning its subject (GPO Stock No. 2915- 0020, $1.50 per copy).

Federal Register. All regulations, procedures, and changes concerning the OSHA act appear in the *Federal Register.* Published six times weekly ($25.00 per year).

"General Industry Guide for Applying Safety and Health Standards, Title 29 CFR 1910" (OSHA 2072). Can be used to identify and locate those standards which apply to specific safety and health hazards as related to the workplace, machines and equipment, materials, employees, power sources, and work processes. Related administrative regulations are also provided. By analyzing this guide and marking the subject areas applicable to specific operations or responsibilities, the user will have a reference to those standards with which he must be concerned (55¢).

Inspection Survey Guide—A Handbook of Guides and References to Safety and Health Standards for Federal Contracts Programs. (Work-place Standards Administration, U.S. Department of Labor). An administrative tool to help safety personnel and the contractor in identifying those standards that will be used to determine compliance with federal standards. It is divided into four parts: Pre-operational, Operational, Appendix, and Subject Index, as follows: Part A—the physical plant; Part B—operational requirements; Appendix—special information relating to federal supply and service contracts; and Subject Index—permits rapid access to pertinent sections of the document. Although this document covers Title 41 CFR Part 50–204—Safety and Health Standards for Federal Supply Contracts, this same title is repeatedly quoted as a regulatory requirement in the OSHA act of 1970. It may almost be considered as setting the standards for the act of 1970 ($2.25).

"Job Safety and Health" (formerly "Safety Standards"). Published by OSHA. Features latest developments, techniques, and programs affecting safety and health in the workplace. Includes listings of the latest insertions appearing in the *Federal Register* concerning the OSHA act, plus other current publications in the fields of safety and health. An excellent tool for keeping abreast of current changes concerning the act ($4.50 per year).

President's Report on Occupational Safety and Health, The, 1972. The first Occupational Safety and Health Report of the President to the Congress contains reports by the Department of Labor and the Department of Health, Education, and Welfare. The 100-page publication reports on research programs, hazard evaluations, compliance operations, the role of the states, and the reporting of work injuries and illnesses ($1.75).

Miscellaneous pamphlets on specific areas (e.g., "The Principles and Techniques of Mechanical Guarding [OSHA 2057]). Contact your nearest OSHA office for the current listing of titles, since these documents come into print at various times and the list will be updated continuously (60¢ per copy).

Subscription Reference Services

Superintendent of Documents
  U.S. Government Printing Office
  Washington, D.C. 20402
All OSHA standards, interpretations, regulations, and procedures are available in easy-to-use looseleaf form, punched for use in a three-ring binder. Changes and additions will be issued for an indefinite period of time.

The following individual volumes may be purchased at the annual rates given in parentheses:

Compliance Manual ($8.00)
Construction Standards ($8.00)
General Industry Standards ($21.00)
Maritime Standards ($6.00)
Other Regulations and Procedures ($5.50)

U.S. Department of Health, Education, and Welfare
Bureau of Product Safety
    5401 Westbard Avenue
    Bethesda, Md. 20016
Write for a complimentary subscription to "Product Safety—NEISS News," an eight-page monthly newsletter published by the Bureau's Injury Data and Control Center (IDCC). The data are collected through a computerized network linking 119 hospital emergency rooms in 30 states to a central computer in Washington, D.C.—the National Electronic Injury Surveillance System (NEISS). Plans are being made for the collection of similar data from physicians and clinics nationwide. This data will provide information on products *involved* in injuries, although not necessarily *causing* injuries, and will subsequently be the basis for establishing target areas of action by the Consumer Product Safety Commission.

National Safety Council
    425 North Michigan Avenue
    Chicago, Ill. 60611
Publishes *OSHA Standards Checklist*, a multivolume checklist of OSHA standards, as well as a comprehensive series of publications on all matters relating to industrial and farm safety. Write for a complete listing of available publications.

Employment Safety and Health Guide
Commerce Clearing House, Inc.
  4025 Peterson Avenue
  Chicago, Ill. 60646
Occupational Safety and Health Reporter
Bureau of National Affairs, Inc.
  Dept. OSHR-554
  1231 25th St. N.W.
  Washington, D.C. 20037

Commercial Publications
Periodicals

American Industrial Hygiene Association Journal (monthly), American Industrial Hygiene Association, 25711 Southfield Road, Southfield, Mich. 48073.

Fire Journal (bi-monthly), National Fire Protection Association, 80 Batterymarch Street, Boston, Mass. 02110.

National Safety News (monthly), National Safety Council, 425 North Michigan Avenue, Chicago, Ill. 60611.

Occupational Hazards (monthly), 614 Superior Avenue, West, Cleveland, Ohio 44113.

Professional Safety (monthly), the official publication of the ASSE, American Society of Safety Engineers, 850 Busse Highway, Park Ridge, Ill. 60068.

General Reference Texts*

Accident Investigation Manual, published by Northwestern University Traffic Institute, 1704 Judson Avenue, Evanston, Ill. 60201.

Accident Prevention Manual for Industrial Opera-

---

* Refer to Appendix A for complete addresses.

tions, F. E. McElroy, ed., published by National Safety Council.

Action Guidelines—National Business Council for Consumer Affairs, published as a public service by Montgomery Ward & Co., Consumer Information Services, 20-N, 535 W. Chicago Avenue, Chicago, Ill. 60607 ($3.00).

Air Sampling Instruments, published by American Conference of Governmental Industrial Hygienists.

Biological Science, American Institute of Biological Sciences, published by Houghton Mifflin Co., 1900 S. Batavia Street, Geneva, Ill. 60134.

Blaster's Handbook, published by E. I. du Pont de Nemours & Co. Inc., Explosives Department, Wilmington, Del. 19801.

"Chemical Safety Data Sheets," published by Manufacturing Chemists Association, 1825 Connecticut Avenue, N.W., Washington, D.C. 20009.

Chemical Safety Supervision, by Robert Guelich, published by Van Nostrand-Reinhold, 450 W. 33 Street, New York, N.Y. 10001.

Chemistry of Powder and Explosives, published by John Wiley & Sons, Inc., One Wiley Drive, Somerset, N.J. 08873.

Consumer Product Safety Act—Law and Explanation, published by Commerce Clearing House, Inc., 4025 Peterson Avenue, Chicago, Ill. 60646.

Dangerous Properties of Industrial Materials, by Irving Sax, published by Van Nostrand-Reinhold, 450 W. 33 Street, New York, N.Y. 10001.

Dictionary of Terms Used in the Safety Profession,

by W. E. Tarrants, published by the American Society of Safety Engineers.

Elevators, by F. A. Annett, published by McGraw-Hill Publ. Co., P.O. Box 509, Hightstown, N.J. 08520.

Environmental Control and Safety Directory, published by A. M. Best Co. Inc., Park Avenue, Morristown, N.J. 07960.

Fire Officers Guide to Dangerous Chemicals, by Charles W. Bahme, published by National Fire Protection Association.

Fire Protection Handbook, published by National Fire Protection Association.

First Aid Textbook, American National Red Cross (consult local chapter for copy).

Fundamentals of Industrial Hygiene, published by National Safety Council.

Guide for Safety in the Chemical Laboratory, Manufacturing Chemists Association, published by D. Van Nostrand & Co., 450 W. 33 Street, New York, N.Y. 10001.

Guidebook to Occupational Safety and Health, published by Commerce Clearing House, Inc., 4025 Peterson Avenue, Chicago, Ill. 60646.

Handbook of Compressed Gases, published by Compressed Gas Association.

Handbook of Dangerous Materials, by Irving Sax, published by Van Nostrand-Reinhold, 450 W. 33 Street, New York, N.Y. 10001.

Handbook of Industrial Loss Prevention, Factory Mutual Engineering Corp., published by McGraw-Hill, P.O. Box 509, Hightstown, N.J. 08520.

Handbook of Laboratory Safety, published by Chemical Rubber Company, 18901 Cranwood Parkway, Cleveland, Ohio 44128.

Handbook of Noise Measurement, published by General Radio Corp., West Concord, Mass. 01742.

Handbook of Rigging, by W. E. Rossnagel, published by McGraw-Hill, P.O. Box 509, Hightstown, N.J. 08520.

Handbook of Solvents, by Leopold Scheflan and Morris Jacobs, published by D. Van Nostrand & Co., 450 W. 33 Street, New York, N.Y. 10001.

Hazardous Materials Guide, published by National Fire Protection Association.

Human Factors Engineering, by E. J. McCormick, published by McGraw-Hill, P.O. Box 509, Hightstown, N.J. 08520.

Industrial Accident Prevention, by H. W. Heinrich, published by McGraw-Hill, P.O. Box 509, Hightstown, N.J. 08520.

"Industrial Safety Data Sheets," published by National Safety Council.

Industrial Hygiene and Toxicology, by F. A. Patty, published by John Wiley and Sons, Inc., One Wiley Drive, Somerset, N.J. 08873, (vol. 1, General Principles; vol. 2, Toxicology).

Industrial Noise Manual, published by American Industrial Hygiene Association, 210 Haddon Avenue, Westmont, N.J. 08108.

Industrial Safety, by R. P. Blake, published by Prentice-Hall, Englewood Cliffs, N.J. 07632.

Industrial Ventilation—A Manual of Recommended Practice, published by American Conference of Governmental Industrial Hygienists.

Inspection Manual, published by National Fire Protection Association.

Manual of Accident Prevention in Construction, published by Association of General Contractors of America, 1957 E Street, N.W., Washington, D.C. 20006.

Modern Chemistry, by H. C. Metcalfe, J. E. Williams, Castka, published by Henry Holt and Co., 383 Madison Avenue, New York, N.Y. 10017.

Modern Safety Practices, by R. DeReamer, published by John Wiley & Sons Inc., One Wiley Drive, Somerset, N.J. 08873.

Motor Fleet Safety Guide, published by National Safety Council.

"National Bureau of Fire Underwriters Pamphlets," published by National Board of Fire Underwriters, American Insurance Association, 85 John Street, New York, N.Y. 10038.

National Electrical Code, published by National Fire Protection Association.

National Fire Codes, (vol. 1, Flammable Liquids; vol. 2, Gases; vol. 3, Combustible Solids, Dusts, and Explosives; vol. 4, Building Construction and Facilities; vol. 5, Electrical; vol. 6, Sprinkler, Fire Pumps, and Water Tanks; vol. 7, Alarm and Special Extinguishing Systems; vol. 8, Portable and Manual Fire Control Equipment; vol. 9, Occupancy Standards and Process Hazards; vol. 10, Transportation), published by National Fire Protection Association.

Physics, A Basic Science, by F. L. Verwiebe, G. E. Van Hooft, B. W. Saxon, published by G. Van Nostrand & Co., 450 West 33 Street, New York, N.Y. 10001.

Principles of Management, by L. J. Kazmier, published by McGraw-Hill, P.O. Box 509, Hightstown, N.J. 08520.

Product Liability and Safety, by G. M. Peters, published by Coiner Publications Ltd., 3066 M Street, N.W., Washington, D.C. 20007.

Property and Casualty Insurance, by Philip Gordis, published by The Rough Notes Co., Inc., 1142 N. Meridian, Indianapolis, Ind. 46204.

Radiological Health Handbook, U.S. Department of Health, Education and Welfare.

Radiological Safety and Major Activities in the Atomic Energy Programs, U.S. Atomic Energy Commission.

Respiratory Protection Manual, American Industrial Hygiene Association, 210 Haddon Avenue, Westmont, N.J. 08108.

Rules and Regulations for Military Explosives and Hazardous Munitions, U.S. Coast Guard, for sale by U.S. Government Printing Office.

Safety and Accident Prevention in Chemical Operations, by Howard H. Fawcett and William S. Wood, published by John Wiley & Sons, Inc., One Wiley Drive, Somerset, N.J. 08873.

Safety Maintenance Directory, published by A. M. Best Co., Inc., Park Ave., Morristown, N.J. 07960.

Safety Management, by Rollin H. Simonds and

John W. Grimaldi, published by Richard D. Irwin, Inc., 1818 Ridge Road, Homewood, Ill. 60430.

Safety Training for the Supervisor, by James E. Gardner, published by Addison-Wesley, Inc., Reading, Mass. 01867.

Selected Bibliography of Reference Materials in Safety Engineering and Related Fields, A, by W. E. Tarrants, published by the American Society of Safety Engineers.

Strength of Materials, by Charles G. Harris, published by American Technical Society, 848 E. 58th Street, Chicago, Ill. 60637.

Supervisor's Safety Manual, published by National Safety Council.

"System Safety Program for Systems and Associated Subsystems and Equipment: Requirements For" (MIL-STD-882), Dept. of Defense, secure copy from Hq. Air Force Systems Command (SCIZS), Andrews AFB, Washington, D.C. 20331

Tariff No. (latest edition), Motor Carriers' Transportation of Explosives and Other Dangerous Articles, American Trucking Association, published by F. G. Freund, 1616 P Street, N.W., Washington, D.C. 20036.

Tariff No. (latest edition), Transportation of Explosives and Other Dangerous Articles, Interstate Commerce Commission, published by T. C. George, 63 Vesey Street, New York, N.Y. 10007.

Threshold Limit Values, published by American Congress of Governmental Industrial Hygienists.

Traffic Engineering Handbook, published by Institute of Traffic Engineers, 1815 N. Fort Meyer Drive, Arlington, Va. 22209.

Unfired Pressure Vessels, published by American Society of Mechanical Engineers, 345 E. 47 Street, New York, N.Y. 10017.